IEEE Radio and Antenna Days of the Indian Ocean (IEEE RADIO 2018)

Extended Papers

IOP Conference Series: Materials Science and Engineering
Volume 524

Wolmar, Mauritius
15 - 18 October 2018

Editor:

Vikass Monebhurrun

ISBN: 978-1-5108-8719-0
ISSN: 1757-8981

Printed from e-media with permission by:

Curran Associates, Inc.
57 Morehouse Lane
Red Hook, NY 12571

Some format issues inherent in the e-media version may also appear in this print version.

Printed by Curran Associates, Inc. (2019)

Institute of Physics
Dirac House, Temple Back
Bristol BS1 6BE UK

Phone: 44 1 17 929 7481
Fax: 44 1 17 920 0979

techtracking@iop.org

Additional copies of this publication are available from:

Curran Associates, Inc.
57 Morehouse Lane
Red Hook, NY 12571 USA
Phone: 845-758-0400
Fax: 845-758-2633
Email: curran@proceedings.com
Web: www.proceedings.com

Table of contents

Volume 524

2018 IEEE Radio and Antenna Days of the Indian Ocean (IEEE RADIO 2018)

15–18 October 2018, Wolmar, Mauritius

Accepted papers received: 1 April 2019

Preface

2018 IEEE Radio and Antenna Days of the Indian Ocean (IEEE RADIO 2018)

Editorial Board Members

Peer review statement

Papers

Validation of electromagnetic numerical simulations through measurements for an efficient deployment of complex systems

A Monorchio, A Corucci and P Usai.....1

Coherent Detection and 3D Tracking Stages of a DVB-T Based Passive Radar for Terrestrial Traffic Monitoring

P J Gómez-del-Hoyo, N del-Rey-Maestre, D Mata-Moya, M P Jarabo-Amores and M C Benito-Ortiz.....5

Fine-tuning calibration of deterministic propagation models in UHF band

G Militaru, D Popescu and L Ichim.....9

Electric field intensity in cross-section of evaporation duct

D G Serebriakov and R R Abdullin.....13

Novel Low THz Frequency Extenders for Vector Network Analyzers

D R Vizard, E G Hoare, M Gashinova and M Cherniakov.....17

Microwave-Photonics Frequency Up-Converter for Telecom and Measurement Equipment

D A Fofanov, T N Bakhvalova, A V Alyoshin, M E Belkin and A S Sigov.....21

Compact directional coupler with five stubs

Denis A. Letavin.....25

RF Single-Pole Double-Throw Switch Based on Two-Port Memistor

J Vavra, J Bajer and D Biolek.....29

Functional MRI and the Lagrangian description of flow

S L Vesely, S R Dolci and C A Dolci.....33

IOP Conf. Series: Materials Science and Engineering **524** (2019) 011001 doi:10.1088/1757-899X/524/1/011001

EDITOR S FOREWORD

The 2018 edition of the IEEE Radio and Antenna Days of the Indian Ocean (IEEE RADIO 2018) international conference was hosted once again in Mauritius, the country where the first three editions of the conference were organized in 2012, 2014 and 2015. Dr. Arjoon Suddhoo, Executive Director of the Mauritius Research Council (MRC), officially opened this sixth edition of the conference that was held from 15th to 18th October 2018 at the Hilton Resort & Spa, Wolmar, Mauritius. The conference was managed by the Radio Society (Mauritius) and financially sponsored by the IEEE Antennas and Propagation Society (AP-S). IEEE RADIO 2018 was technically co-sponsored by IEEE Region 8 and Union Radio Scientifique Internationale (URSI). Researchers from CentraleSupélec, France, University of Mauritius and Université de La Réunion formed the local organizing committee.

Participants from over twenty different countries spanning all five continents attended IEEE RADIO 2018. The conference featured fourteen regular sessions covering diverse topics such as antenna design, devices and circuits, wave propagation and scattering, wireless communication systems, computational electromagnetics, and biological effects of electromagnetic fields. A tutorial on computational electromagnetics was proposed on Monday morning. During the opening ceremony, the IEEE Ulrich L. Rohde Humanitarian Technical Field Project Award was presented to IEEE Special Interest Group on Humanitarian Technology (SIGHT) Indian Ocean for the Agriculture, Climate and Technology in Indian Ocean Network (ACTION) project. Moreover, a SIGHT Event was organized on Monday afternoon during which representatives from different IEEE SIGHT Groups presented their humanitarian activities and projects. Three internationally recognized scientists delivered invited talks during the conference.

A jury evaluated the oral presentations of students and young scientists. The Best Young Scientist less than 36 years old received a shield and a cash prize of 300 Euros. The first, second and third Best Student Papers were awarded shields and cash prizes of 300 Euros, 200 Euros and 100 Euros, respectively. The inaugural Industrial Engineering Paper Award, consisting of a shield and a cash prize of 300 Euros, was presented to the author of the paper presenting the most innovative and significant contribution in terms of practical industrial engineering related to the fields of antennas, electromagnetics and propagation.

The 2-page short papers presented during the conference have already been published on the IEEE Xplore Digital Library. After the conference, the authors of the accepted and presented papers were invited to submit an extended version (limited to a maximum of 4 pages) for publication as a volume in the IoP Conference Series: Materials and Engineering (MSE). Eighteen papers were submitted. After the peer-review process by the distinguished members of the Scientific Committee, only nine papers were accepted and published in this volume of IoP Conference Series: MSE.

Editor
Vikass Monebhurrun

IOP Conf. Series: Materials Science and Engineering **524** (2019) 011002 doi:10.1088/1757-899X/524/1/011002

IOP Conf. Series: Materials Science and Engineering **524** (2019) 011003 doi:10.1088/1757-899X/524/1/011003

Peer review statement

All papers published in this volume of *IOP Conference Series: Materials Science and Engineering* have been peer reviewed through processes administered by the proceedings Editors. Reviews were conducted by expert referees to the professional and scientific standards expected of a proceedings journal published by IOP Publishing.

Validation of electromagnetic numerical simulations through measurements for an efficient deployment of complex systems

A Monorchio[1,2], A Corucci[2] and P Usai[3]

[1]University of Pisa Pisa, Italy
[2]RaSS Lab C.N.I.T. Pisa, Italy
[3]Free Space srl Pisa, Italy

a.monorchio@ieee.org, a.corucci@cnit.it, pierpaolo.usai@free-space.it

Abstract. A proper use of electromagnetic numerical simulations can lead to an efficient equipment integration and consequent successful deployment of complex systems and stations. In this paper, the level of accuracy attainable with modern numerical methods is discussed as a consequence of measurement campaign performed in complex systems and environments.

1. Introduction

Electromagnetic numerical simulations have become a standard tool for the design of RF and microwave devices and subsystems. The growing maturity of such techniques, together with available high power computing capabilities, allowed us to go several steps further in the design stage of complex systems so that a proper use of such techniques can lead to an efficient equipment integration and consequent successful deployment of complex systems and stations. In particular, the final goal of an interference-free system, maintaining, at the same time, the desired performance of each subsystem to be integrated, is a key issue.

The enormous recent development of numerical methods avoids costly trial and error procedures based on post factum verifications, providing a remarkable benefit: different engineered solutions can be tested in advance, with the great advantage of an early knowledge of the attainable performance; as a result, technical specifications of the overall systems, issued by buyers and stakeholders, can now be tailored to the available and cost-efficient technologies.

The above-mentioned issues pose different questions about a major concern, i.e., how far the numerical accuracy must be pushed to obtain meaningful results for a correct final deployment of complex systems/stations. Some results will be presented based on authors' experience on numerical methods, and compared with real measurements obtained in practical cases, regarding satellite ground stations, military base with the simultaneous presence of communication and electronic warfare systems, top side arrangement of ships.

2. Early stage design: simulation methods

2.1. Setting up scenery

The first step to be taken before running any simulation is the construction of the scenario where sources and victims operate. Generally, we start from a model of the complex scenario (i.e. ship, airplane, helicopter) built for purposes other than those of electromagnetic analysis. Therefore, the first effort to be made is to adapt the mechanical model to the needs of the chosen numerical methods. The adaptation

IOP Conf. Series: Materials Science and Engineering **524** (2019) 012001 doi:10.1088/1757-899X/524/1/012001

takes place eliminating from the mechanical model all the details irrelevant for electromagnetic purposes and building a mesh of the object that respects the constraints of the chosen model. After having defined the geometry of the scenario it is necessary to characterize its parts with the type of material that compose them through its electrical properties. These properties can be found in literature or directly measured if unknown.

2.2. Characterization of materials

The description of the materials from the point of view of their electromagnetic properties is as important as the geometric description. The designer can face two situations: either the measurement data of the electrical properties of the materials in the band of interest are available or it is necessary to refer to data present in the literature. In the former case, the designer will have to consider the correctness of the method used to estimate the parameters to be employed in his model, whereas in the latter case, not all materials may be available in open literature so one may need to refer to similar experiences for the model to be realistic. Furthermore, it is not possible to characterize a material on a broad frequency band by using a single method and a single measurement setup. In the case of a waveguide measurement at a given band up to GHz, there will certainly be a need for more setups, i.e., as many setups as the number of waveguides necessary to cover the band of interest. If we are also interested in two polarizations, we must also multiply by two the measurement effort.

Among the methods based on the inversion of the Fresnel formulas expressing the reflection and transmission coefficients at the interface of a material layer by resorting to the wave impedance of the medium and its index of refraction, the method of Nicolson-Ross -Weir (NRW) is certainly one of the most used [1], cited and modified by engineers [2].

To characterize a material in the 4-18 GHz radar band, it is necessary to use a set of waveguides of different sizes in order to cover the whole band. Table 1 shows a possible choice of three waveguides for the band of interest and indicates the name, the dimensions of the long (a) and short (b) sides and the band covered by each guide: the waveguide with larger geometrical dimensions (WR187) corresponds to the narrower frequency band, whereas the smaller one (WR62) has the wider band.

Table 1. Size and frequency band of three waveguides used to characterize a plastic sheet

Wave Guide	Parameters a[mm] x b[mm]	Band [GHz]
WR187	47.548 x 22.148	3.95 - 5.85
WR90	22.86 x 10.16	8.20 - 12.40
WR62	15.798 x 7.8994	12.40 - 18

Table 2. Electric parameters of a plastic sheet estimated by using the Nicholson-Ross-Weir method

Properties	Plastic sheet
Dielectric relat. permittivity	3.0 - 3.5
Magnetic relat. permeability	0.9 - 1.1
Conductivity [S/m]	0.02 – 0.2

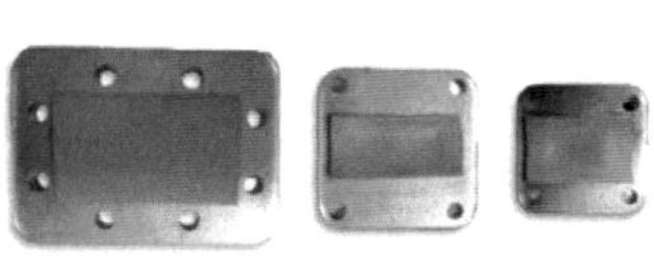

Figure 1. The waveguides WR187, WR90 and WR62.

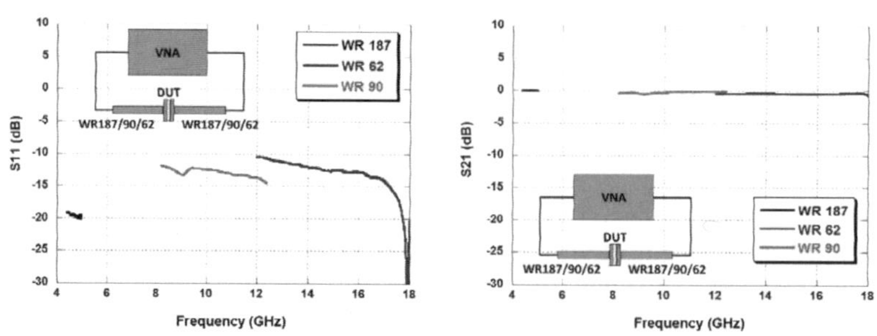

Figure 2. Measured S11 and S21 parameters (setup shown in the inset) of a 0.65 mm thick plastic sheet.

It is therefore clear that investigating the lower portion of the band requires a greater number of guides than the upper portion 12-18 GHz, which is covered by the WR62 alone. Fig. 1 shows the three waveguides with a sample of plastic, a low loss dielectric material, and Fig. 2 the S11 and S21 parameters measured by using the setup shown in the inset of the figure. Table 2 shows the estimated dielectric permittivity, magnetic permeability and conductivity of the plastic sheet (minimum and maximum values).

2.3. Numerical simulations

Commercial software for calculating low-frequency electromagnetic propagation is based on numerical methods such as Method of Moments (MoM), Integral Equation Method and Finite Element Method or Finite Difference Time Domain. These methods estimate the exact solution of the Maxwell equations, but they need to discretize the objects of the scenario by a mesh whose size is of the order of one tenth of the wavelength. This causes an increase in the simulation time, more evident when the frequency increases, that has been partially overcome by the implementation of speed-up algorithms, such as the Multi-Level Fast Multipole Method or the Characteristic Basis Function method [3] for the MoM. Unfortunately, these full-wave EM computational techniques have proved unsuitable for simulating environments that exceed 100 wavelengths [4]. In these cases, the propagation of wireless electromagnetic signals is studied by using high-frequency techniques based on Geometrical Optics, the Geometric Theory of Diffraction and its uniform extension such as the Uniform Theory of Diffraction, because they show some advantages: no constraints on the size of the mesh, ray tracing algorithms can be used to compute the wave propagation in the inner side and the attenuations that rays suffer scenarios (reflections, transmissions and diffractions) are easily evaluated [5][6].

3. Measurements in complex systems and environments

In a complex system or platform, measurement campaigns are needed to verify both EMC issues (as the RADHAZ, HERE, HERO, HERF) and personnel exposure (HERP). It is necessary to compare the level of the electric field emitted by the sources (i.e. antennas) with the maximum level reported in the regulations (i.e. MIL-STD-461/464, STANAG 2345, ICNIRP). Considering that measurements can be performed both by broadband instrumentation (isotropic field sensors for E and H fields) and by narrow band instrumentation (by means of spectrum analyzer) they can be very expensive in terms of time and cost. To make the measurement process more efficient, we propose a working method based on the synergy of measures and simulations. Indeed, thanks to the simulations, the most critical measuring points can be identified and therefore limiting the measurements to only these points. Once the agreement between measurements and simulations has been verified at these points, simulations can be used instead of measurements to evaluate the electric field at any other point. Fig. 3 and Fig. 4 show some photos of measurements carried out for HERP and HERE purposes and the related simulated scenarios. Table 3 shows the comparisons between the measured and simulated electric field values due to a radar antenna and a VHF communication antenna.

It is worth mentioning that similar accurate results can also be obtained in terms of other meaningful parameters to system engineers, such as the delay spread or angle of arrival ([7][8]).

This kind of procedure (simulations and measurements both broadband and narrowband) has been carried out extensively during several projects in recent years ([6] [8] [9]). Therefore, it is possible to state that, if the scenario and the sources are faithfully characterized from an electromagnetic point of view, in rural scenarios (having large size and few obstacles) the error between simulations and measurements is about ± 1.5dB, while in complex scenarios (small size and many details, for example aboard ships) the error is about ±3dB.

4. Conclusion

In this paper, a discussion about the simulation accuracy attainable by current electromagnetic methods has been conducted with particular reference to the complexity of the system under investigation, [10]. Measurements performed after the deployment of the system show that a sufficient

IOP Conf. Series: Materials Science and Engineering **524** (2019) 012001 doi:10.1088/1757-899X/524/1/012001

accuracy can be obtained if the initial model contains the main features contributing to the propagation phenomena: obstacles and their material electric properties; exhaustive description of sources. Such a high degree of confidence allows designers to carefully plan in advance the layout of ICT equipment and subsystems for maximum performance [11] and interference-free functioning.

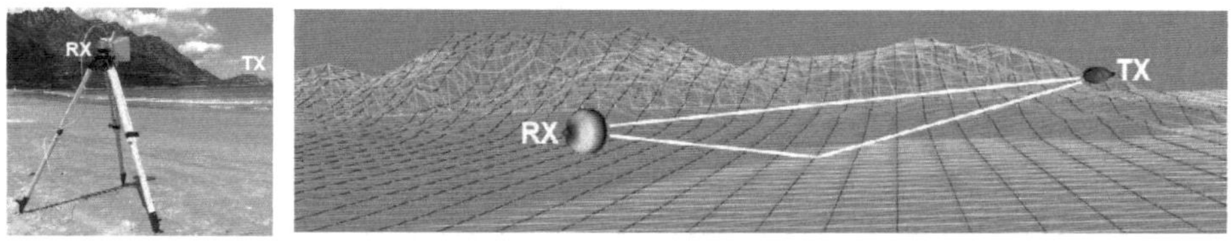

Figure 3. Measurement and simulation for radar antenna.

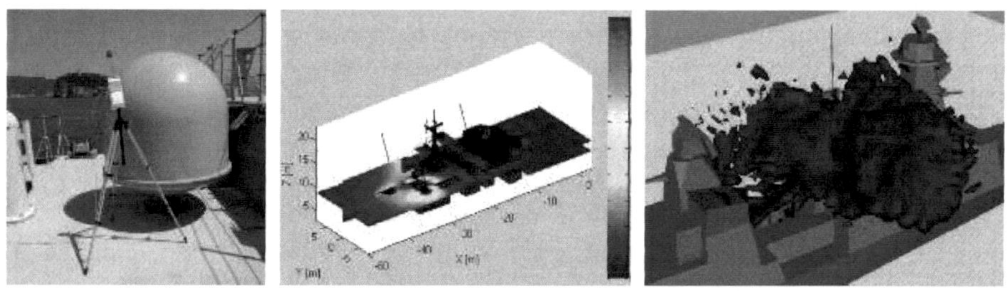

Figure 4. Measurement and simulation for VHF antenna.

Table 3. Measurement and simulation comparison related to scenarios of Figure 4 and 5

Scenario	Electric Field [V/m]	
	Measured	Simulated
Radar	211	215
VHF	70	80

References

[1] Luukkonen O, Maslovski SI, Tretyakov SA 2011 *IEEE Antennas and Wireless Propagation Letters* vol 10 pp 1295–8

[2] Degiorgi M, Costa F, Fontana N, Usai P and Monorchio A 2016 *IEEE Int. Symp. on Antennas and Propagation (Fajardo, Puerto Rico)* pp 1989-1990

[3] Lucente E, Monorchio A and Mittra R 2008 *IEEE Transaction Antennas and Propagation* vol 56/4 pp 999–1007

[4] Stara R et al., 2013 *Progress In Electromagnetics Research M* vol. 29 pp 121–136

[5] A. Corucci, Usai P, Monorchio A and Manara G 2014 *Computational Electromagnetics: Recent Advances and Engineering Applications* ed R Mittra (NY: Springer) chapter 17 pp 575–618

[6] Tiberi G, Bertini S, Malik W Q, Monorchio A, Edwards D J and Manara G 2009 *IEEE Trans. Antennas Propag.* vol. 57 pp. 777–785

[7] Sfrecola M, Corucci A and Monorchio A 2016 *IEEE Int. Symp. on Antennas and Propagation (Fajardo, Puerto Rico)* pp 1263-1264

[8] Usai P, Corucci A, Gligorevic S and Monorchio A 2011 *5th European Conf. on Antennas and Propagation EUCAP* pp 1634-1638

[9] Monorchio A, Corucci A, Bertoneri M and Angeloni M 2012 *Atti del convegno Campi elettromagnetici e innovazione tecnologica in ambito Difesa, Industria e Ricerca (Pisa, Italy)* vol 1 pp 32-34 (ISBN 978-88-6741-169-6)

[10] Monorchio A, Corucci A and Usai P 2018 *IEEE RADIO (Mauritius)* pp 1-2

[11] Mittra R, Li C, Sharawi M S and Monorchio A 2016 *10th European Conf. on Antennas and Propagation EUCAP* pp 1-2

IOP Conf. Series: Materials Science and Engineering **524** (2019) 012002 doi:10.1088/1757-899X/524/1/012002

Coherent Detection and 3D Tracking Stages of a DVB-T Based Passive Radar for Terrestrial Traffic Monitoring

P J Gómez-del-Hoyo, N del-Rey-Maestre, D Mata-Moya, M P Jarabo-Amores, M C Benito-Ortiz

Department of Signal Theory and Communications, Superior Polytechnic School, University of Alcalá, Alcalá de Henares, Madrid, Spain

{pedrojose.gomez; nerea.rey; david.mata; mpilar.jarabo;cortes.benito}@uah.es

Abstract. This work tackles terrestrial traffic monitoring using a Passive Radar System (PRS) characterized by a surveillance channel processing composed of a Neyman-Pearson (NP) coherent detector and three-dimensional tracker implementation. The NP detector is a parametric approach that requires the prior knowledge of target and bistatic clutter statistical models at communication systems frequencies that fit with real data. The proposed tracker provides three-dimensional (range, azimuth and speed) trajectories declaration exploiting the capabilities of beamforming techniques based on a uniform linear array as surveillance antenna system. The considered solution is validated with real data acquired by a PRS in a semi-urban scenario.

1. Introduction

Non-collaborative target detection and tracking are a challenging problem of great interest. Bistatic Passive Radar Systems, PRS, are being extensively developed as an alternative solution to active ones, due to advantages such as low development, implementation and maintenance cost, small size, low weight, low probability of interception and avoidance electromagnetic compatibility or environment impact problems. All these features are a result of an opportunistic use of the transmissions of other non-cooperative radiocommunication services present in their surrounding area as Illuminators of Opportunity (IoO).

PRSs scheme is usually composed of two channels: the reference channel for acquiring IoO signal and the surveillance channel for capturing the target echoes. The two acquired signals are coherent processed in intervals (CPI, Coherent Processing Interval) consisting in the correlation of time delayed and Doppler-shifted copies of the selected reference signal with target echoes to generate Range-Doppler Surfaces (RDSs) using the Cross-Ambiguity Function. PRSs performances are strongly determined by the IoO properties and the radar scenario geometry.

The radar detector has to decide between target absence (null hypothesis, H_0) and target presence (alternative hypothesis, H_1). The most extended detector criterion in radar applications is the Neyman-Pearson (NP) detector, which maximizes the Probability of Detection (P_D) maintaining the Probability of False Alarm (P_{FA}) lower than or equal to a given value [1]. If $\tilde{\mathbf{z}}$ is the complex observation vector provided by the radar receiver and $f(\tilde{\mathbf{z}}|H_0)$ and $f(\tilde{\mathbf{z}}|H_1)$ are the detection problem likelihood functions under both hypotheses, a possible implementation of the NP detector consists in comparing the Likelihood Ratio (LR) to a detection threshold estimated according to P_{FA} requirements [2]. This approach requires a complete statistical characterization under both hypothesis and significant detection losses arise when the statistical properties of the observation vectors under both hypotheses vary from

IOP Conf. Series: Materials Science and Engineering **524** (2019) 012002 doi:10.1088/1757-899X/524/1/012002

those assumed in the LR detector design [3]. There are not available too many studies of bistatic interference or clutter at communication systems frequencies in the literature. In [4], a two-dimensional (2D) LR detector is designed for a PRS based on Digital Video Broadcasting-Terrestrial (DVB-T) after proving that Gaussian and mixture of Gaussians statistical models fit with real data.

The three-dimensional (3D) passive radar detection and tracking consider target bistatic range, Doppler shift and azimuth information. Antenna array and digital beamforming techniques are required in the surveillance channel to provide the angular discrimination capability. The 3D-detector is usually based on two-stage processing scheme: the first one consists in a 2D-detector applied to the RDSs associated with each array element and a combination of the detection performances, and the second one is focused on Direction-of-Arrival (DoA) estimation techniques. In [5], Constant False Alarm Rate (CFAR) techniques (that can lead to suboptimal solutions) exploiting signals acquired by non-Uniform Linear Array (non-ULA), a 2D-tracker and a high resolution DoA estimator are proposed.

In this paper, an enhanced 3D-detector for IDEPAR (Improved Detection techniques for Passive Radars), a passive radar demonstrator, is proposed. IDEPAR is a multi-channel DVB-T PRS designed and developed for terrestrial targets [6], and its capabilities were evaluated in [4]–[7]. The antenna system of its surveillance channel is composed of an array of commercial antennas fulfilling the IDEPAR design requirement of using Commercial Off-The-Shelf (COTS) components to the largest possible extent. To improve the PRS performance, this work considers the LR detector implementation, a DoA estimator based on ULA and MUSIC algorithm [8], and 3D-tracker using all the target information to reduce anomalous trajectories.

2. 3D Passive Radar Detector and Tracker

For the formulation of the LR detector (1), where i denotes the ith-element of the surveillance antenna array composed of N elements, RDSs provided by each array element are used as the observation space and $f(\tilde{\mathbf{z}}|H_0)$ is defined by the ground clutter models proposed in [4] for the different areas of the RDSs. These models present small variations along time that can be characterized. A Gaussian target is assumed for $f(\tilde{\mathbf{z}}|H_1)$. The threshold ($\eta_{LR}$) is fixed attending to a given P_{FA}.

$$\Lambda^{i}(\tilde{\mathbf{z}}) = \frac{f(\tilde{\mathbf{z}}|H_1)}{f(\tilde{\mathbf{z}}|H_0)} \underset{H_0}{\overset{H_1}{\gtrless}} \eta_{LR}^{i}\,(P_{FA}) \qquad i = 1, \dots, N \tag{1}$$

The N LR detector outputs are proposed to be combined by an OR-logic operation resulting a detection matrix indicating the range-Doppler cells where targets can be present, so this stage provides target 2D-dimensional parametrization. As η_{LR}^{i} is fixed to fulfill the P_{FA} requirements of only the ith-antenna, an increase of the final P_{FA} after the OR-logic combination is expected. However, once a target detection is loss in this step, it will never be considered in following steps. In addition, P_{FA} performance can be improved in the tracking stage.

The use of beamforming techniques provides the third dimension of the target characteristics space estimating its azimuth. MUSIC algorithm is applied to the range-Doppler cells where targets are declared, using N-dimension frequency domain snapshots composed of the corresponding samples of the RDSs generated for each array element. MUSIC estimator is widely used because this solution is computationally simpler than the Maximum Likelihood estimator providing good performance capability [8]. MUSIC algorithm finds the directions vectors that are orthogonal to a noise subspace exploiting the information of the eigenvalues and eigenvectors of the spatial covariance matrix.

The association of the temporal detections from the same target is carried out by a 3D-tracker based on a 3D Kalman filter. The considered solution is an updated version of the 2D-tracker presented in [5]. The target manoeuvres in the domain under study are described by equations (2-3) of dynamic model. The target movement in the bistatic range dimension follows a basic accelerate movement where the acceleration variation is modelled as white noise, $\omega_R(k)$. The instantaneous bistatic range position of the target is denoted as $R_b(k)$, and its velocity and acceleration as v_{Rb} and a_{Rb} respectively. The azimuth displacement, $\phi(k)$, is modelled as a lineal movement with additive white noise to represent the azimuth

IOP Conf. Series: Materials Science and Engineering **524** (2019) 012002 doi:10.1088/1757-899X/524/1/012002

rate variations, $\omega_{\phi}(k)$. v_{ϕ} represents the change rate in this dimension. In both equations, the time of the CPI is denoted as T_{int}.

$$R_b(k) = R_b(k-1) + v_{Rb}(k-1) \cdot T_{int} + \frac{T_{int}^2}{2a_{Rb}(k-1)} + \omega_R(k); \quad (2)$$

$$\phi(k) = \phi(k-1) + v_{\phi}(k-1) \cdot T_{int} + \omega_{\phi}(k) \quad (3)$$

3. Evaluation with IDEPAR data

A real radar scenario, in the nearby area of the Polytechnic School of University of Alcalá (Spain), is considered. IDEPAR (Figure 2), DVB-T based PRS, is located at the rooftop terrace of the Polytechnic School. This semi-urban scenario is described in [4]-[7]. The considered 3D radar signal processing scheme of IDEPAR is depicted in Figure 2.

Figure 1. IDEPAR PRS

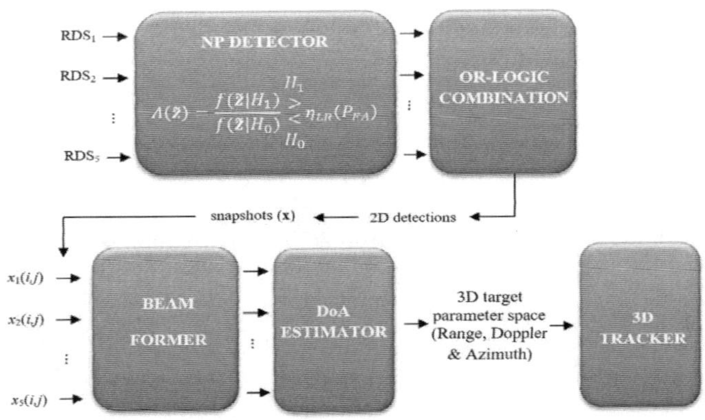

Figure 2. 3D passive radar detector and tracker scheme

The acquisition parameters are:58-th DVB-T channel (f_c=770 MHz); Signal bandwidth of 8 MHz; Coherent processing interval of T_{int}=250 ms; RDS size of 1000 range bins (corresponding to a coverage distance of 9.45 km) and 401 Doppler shifts (f_{DOP} = [-799.744, ...,799.744] Hz); An acquisition of 20 seconds was recorded and 80 RDSs are generated. IDEPAR surveillance antenna system is a ULA of Televés 4G NOVA antennas (Figure 3): a commercial UHF log-periodic antenna designed in microstrip technology covering IDEPAR frequency band with a 3dB azimuth beamwidth of approximately 60° and a directivity of 7.98 dBi for 770 MHz. The inter-element spacing of ULA is 315 mm, so beam patterns for steering angles ∈ [-13.7°, 13.7°] will not present grating lobes (Figure 4).

In [7], the coherent detection stage is evaluated. The implementation of the LR detector clearly reduces the interference. In Figure 6, the superimposition of the detections declared at the output of the OR-logic combinatory in the whole acquisition time is presented. The false detections associated with the neighborhood metal buildings will be removed by the 3D tracker because is not being able to establish any correspondence with any terrestrial target movement dynamics. The trajectories declared by the 3D tracker (Figure 10) confirm the suitability of the considered solution.

4. Conclusions

A coherent detection and 3D tracking solutions are proposed for terrestrial traffic monitoring using PRS based on DVB-T as IoO. A two-stage processing scheme is designed: the first stage consists in implementing a LR coherent detector to each surveillance acquisition channels and a combination of the outputs with an OR-logic operation; in the second stage a DoA estimator based on digital beamforming techniques and MUSIC algorithm is performed. UHF terrestrial bistatic clutter models are considered for approximating the NP detector and providing the best detection capability. Surveillance ULA completes the 3D target features extraction and allows the 3D tracking. Real data acquired by IDEPAR (DVB-T PRS) in a semi-urban scenario are used to confirm the suitability of the presented radar processing approach.

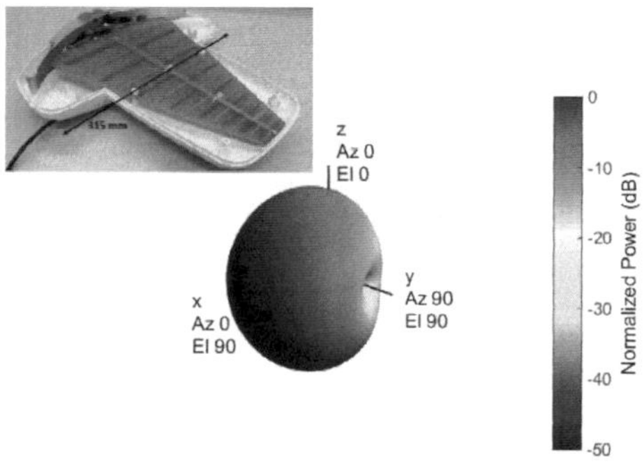

Figure 3. Televes 4G NOVA radiation pattern

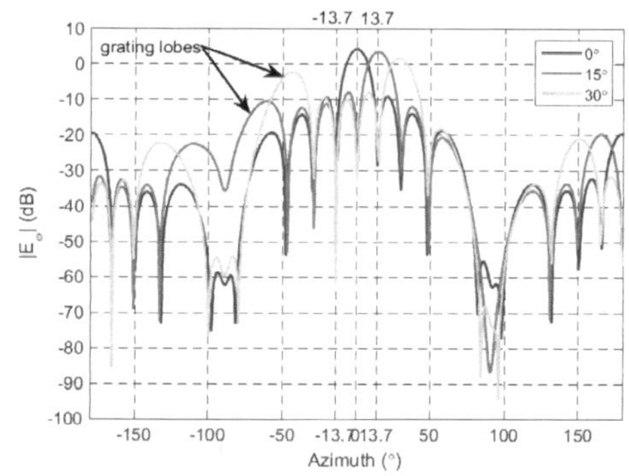

Figure 4. ULA radiation pattern synthesis

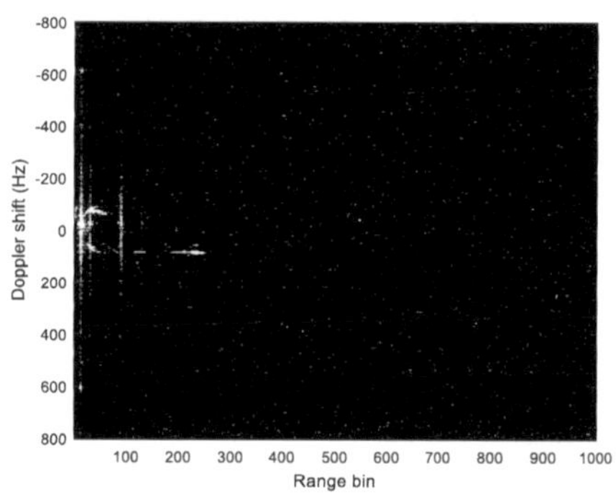

Figure 5. Superimpositon of 2D detections

Figure 6. Terrestrial Target Tracking Performance

5. References

[1] Neyman J and Pearson E 1933 On the problem of the most efficient test of statistical hypotheses *Philosophical Trans. of the Royal Society of London* vol A 231 no 9 pp 289–337

[2] Trees H V 2013 *Detection, estimation, and modulation theory* (Wiley)

[3] di Vito A. and Naldi M. 1999 Robustness of the likelihood ratio detector for moderately fluctuating radar targets *IEE Proceedings on Radar, Sonar and Navigation* vol 146 no 2 pp 107–112

[4] del-Rey-Maestre Net all 2018 Machine Learning Techniques for Coherent CFAR Detection Based on Statistical Modeling of UHF Passive Ground Clutter *IEEE J. of Selected Topics in Signal Processing* vol 12 no 1 pp 104-118

[5] del-Rey-Maestre N et all 2017 Passive Radar Array Processing with Non-Uniform Linear Arrays for Ground Target's Detection and Localization *Remote sensing* vol 9 no 7 pp 1-29.

[6] Jarabo-Amores M P et all 2016 IDEPAR: a multichannel digital video broadcasting-terrestrial passive radar technological demonstrator in terrestrial radar scenarios *IET* Radar, Sonar and Navigation pp 1–9

[7] Gómez-del-Hoyo P J et all 2018 Terrestrial Targets Tracking with Passive Radar System Using Neyman-Pearson Detector *2018 IEEE Radio and Antenna Days of the Indian Ocean* pp 1-2

[8] Yu T and Hansen J H L 2010 Automatic beamforming for blind extraction of speech from music environment using variance of spectral flux-inspired criterion *IEEE Journal of Selected Topics in Signal Processing* vol 4 no 5 pp 785–797

Acknowledgments

This work has been partially funded by the Spanish "Ministerio de Economía, Industria y Competitividad", under project TEC2015-71148-R, the University of Alcalá, under project CCG2017/EXP056

Fine-tuning calibration of deterministic propagation models in UHF band

G Militaru, D Popescu and L Ichim

Faculty of Automatic Control and Computer Science, University Politehnica of Bucharest, Bucharest, Romania

Abstract. A global characterization of the propagation behaviour that will allow a general approach of the Radio Frequency (RF) prediction is difficult to achieve. In this paper, based on field strength values collected in typical rural environment, the authors are providing their findings after the calibration of a deterministic propagation model in Ultra High Frequency (UHF) frequency band. An evaluation of the error sources, and methods to validate RF antenna pattern, are described. The proposed methodology was used to obtain the measurements that allow the correlation with prediction data.

1. Introduction

Due to the complexity of phenomena and considering the evolution of new technologies and the need of cost-effective radio design solutions and of more reliable prediction methods, the modeling of the interaction between the radio waves and the propagation environment requires permanent interest. In addition to well documented propagation studies [1] - [5] the researchers are interested in analyzing various propagation models that provide the best accuracy in predicting RF (Radio Frequency) field strengths for different environments, technologies and frequency bands. A wide variety of RF prediction software tools offer to the RF design engineer a wide range of propagation models that can be used to simulate and predict the radio-link behavior and the quality of the connection. Several works analyzed the behavior of well-known propagation models in comparison with measurements sets [6] - [8]. Whether a geometrical (deterministic), statistical (empirical) and mixed propagation model are available with RF software tools, selecting the right propagation model is quite challenging. Often, RF researchers make several drive tests to collect data that will allow the tuning of the best propagation model. For the geometrical propagation model, the main attenuation factors and losses that impact the signal of interest are described hereafter: in [9] - the free-space attenuation, in [10] - the diffraction loss, in [11], [12] - the gas attenuation, and in [13] - the rain attenuation.

In this paper, the novelty of the author's approach is that the propagation model has been fine-tuned based on real measurements collected during two drive tests. The users can determine the optimum values to be defined as the default settings of the calibrated propagation model. The results of the correlation between simulation and measurements demonstrate the practicality of the proposed methodology and prove that the theoretical prediction is aligned with the measurements.

2. Methodology

The free-space attenuation expressed in dB is calculated as $L_{fs}=20 \cdot \log\left(\frac{4\pi d}{\lambda}\right)$,where λ represents the wavelength and d is the radio path length [9]. The diffraction loss happens when the Fresnel ellipsoid is obstructed by obstacles, buildings, forest. A good approximation to the knife-edge diffraction loss is (1) [10]:

$$L_d = 6.9 + 20log\left[(\nu - 0.1) + \sqrt{1 + (\nu - 0.1)^2}\right] \quad (1)$$

where ν is the Fresnel-Kirchhoff diffraction parameter $\nu = h \cdot \sqrt{\frac{2(d_1+d_2)}{\lambda \cdot d_1 \cdot d_2}}$, h represents the height of the obstruction relative to the Line Of Sight (LOS) and d_1, d_2 are the radio path lengths in meters from each of the end points to the diffraction edge.

The gas attenuation is calculated according to recommendation ITU-R P.676-3 [11] as $L_{gas} = \gamma(f, T, \rho, p)d$, where d is the distance from transmitter location. The coefficient γ depends on the water vapor density ρ (fixed), the temperature T (fixed), the atmospheric pressure p (fixed) and the frequency f (the only variable) [12].

The rain attenuation is calculated following the recommendations ITU-R PN.837-1/838 [13] as $L_{rain} = \gamma(f, R)d$, where γ gets a value depending on the frequency f between 1 and 400 GHz and on a rainfall intensity R, depending on the location on Earth and on time variability. All of the above losses and attenuations are providing too optimistic field strength values. In [14] an additional correction term is proposed, called subpath attenuation (2):

$$L_{sl} = 20\log(7500d) - 20\log(\pi h_1 h_2 f) \quad (2)$$

where d is the distance to transmitter, h_1 and h_2 are respectively the transmitter and receiver antenna heights and f is the frequency.

A different option for increasing the accuracy of the prediction is given by the ground occupancy layer in which the signal of interest is traveling from transmitter to receiver, called ground occupancy correction. The authors of this study are focusing in finding the values of the parameters characterizing the clutter, to optimize the accuracy of the prediction.

This paper presents the parameters of the ground occupancy layer that characterizes the propagation environment, obtained through the correlation of field strength measurements and prediction values calculated using a deterministic propagation model - ITU-R P.525/526 [15] - incorporated in ICS Telecom, an RF prediction software tool produced by the French company, ATDI. The main steps the authors performed to achieve the results presented in this paper, are described hereafter:

1) *Setting the computer aided simulation environment:* a)Setting up ICS Telecom, produced by ATDI and the cartographic dataset - The Digital Terrain Model (DTM) and ground occupancy layer (clutter), b)Pre-configuration of the propagation model - Free space loss (ITU-R P.525/526) + Diffraction loss (ITU-R P.526) + Sub-path loss, c) Equipment setup – CalAmp Radio, Viper SC+ 400 configured at 1 W ERP @ 446 MHz with a 0 dBi omnidirectional antenna placed at 1 m above ground level.

2) *Error source analysis and corrections:* a) The quality of the cartography used and its resolution is directly influencing the simulation results. The cartographic dataset is a medium resolution dataset formed by the DTM, the image layer and the clutter layer with the following clutter classes: open/rural, village, suburban, urban, forest, parks, water, corn fields, common crops. b) The accuracy of the Global Positioning System (GPS) used in the signal acquisition – inadequate GPS measurements taken with less than 6 satellites connection have to be eliminated. c) The fast/slow fading - Lee's Criteria [15] has to be applied, to achieve a good quality of the measurements. d) The technical parameters of the RF equipment and of the Tx antenna - Validating the parameters with measurements in LOS areas and within the reliable distance (figure 1) is mandatory.

3) *Drive tests:* Two drive tests conducted in the surrounding of the Meteorological Station connected to the RF transmitter allowed the authors to better analyze the clutter impact on the RF propagation.

4) *The correlation of the measurements and the simulation.* This allowed the authors to determine, in ICS Telecom, the standard deviation (SD) and average error (AE) of the selected propagation model with the default settings. The correlation has been analyzed in three different scenarios: scenario 1 - The default propagation model, scenario 2 - scenario 1 with default clutter type and clutter height, scenario 3 - scenario 2 with fine-tuned clutter type, clutter height and clutter attenuation.

5) *Tuning the calibration model parameter - optimize the SD and AE.* The steps described in figure 2 were applied to determine acceptable values for the two parameters (SD and AE) and to provide RF planners with reliable values for the ground occupancy parameters that will ensure the best accuracy for further simulations.

IOP Conf. Series: Materials Science and Engineering **524** (2019) 012003 doi:10.1088/1757-899X/524/1/012003

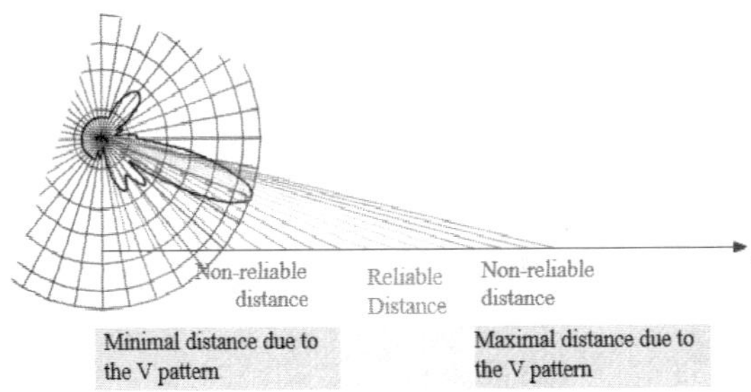

Figure 1. Main antenna lobe and reliable distance for validation.

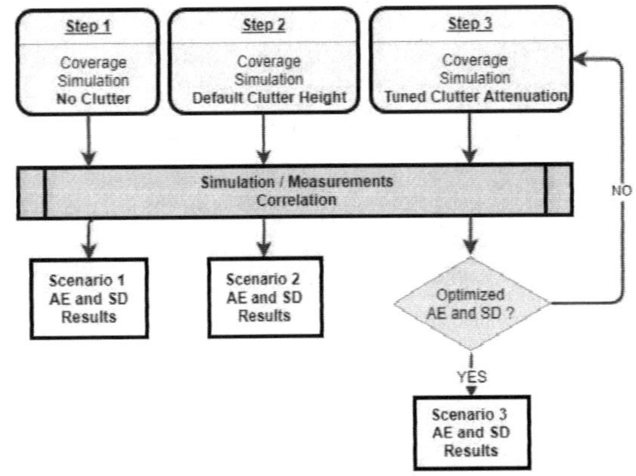

Figure 2. Calibration Methodology.

Based on the existing measurements, the authors' objective is to obtain better standard deviation and average error by applying different values of the parameters characterizing the ground occupancy (clutter height and clutter attenuation). The calibration options available in ICS Telecom have been used and the determined parameters ensured a significant increase in accuracy.

3. Results and discussions

Continuing the work in [16], the authors of this paper explained in details the methodology applied for the proposed process of calibrating the deterministic propagation model that allows the calculation of the model accuracy parameters, presented in table 1. Ground occupancy parameters obtained after the calibration process are presented in table 2. The coverage maps presented in figure 3 are allowing the authors formulating the following conclusions: 1) The impact of the clutter layer (ground occupancy layer) on the RF coverage prediction is obviously noticed by comparing scenario 1 coverage map and scenario 2 coverage map. 2) When introducing attenuation factors for each clutter type, we notice significant differences inside the areas of each clutter type (scenario 2 vs. scenario 3). 3) It can be noted that once the full parameters characterizing the ground occupancy layer are defined (scenario 3), the coverage map is more accurate and the total covered surface is decreased (due to the attenuations and reflection losses introduced by the clutter types). As it can be seen in figure 4, the selected deterministic propagation model and the medium resolution cartographic dataset that is including the clutter layer characterized by the parameters obtained after the calibration process are allowing RF design engineers to get an accuracy of prediction with SD less than 2 dB with an AE of approximately -1 dB.

4. Conclusions

The results are subject to the limitations of this study considering the input data. An increased number of measurement points in various terrain and ground occupancy conditions can bring more consistent conclusions since the selected scenario to run this study does not exhaustively cover all conditions.

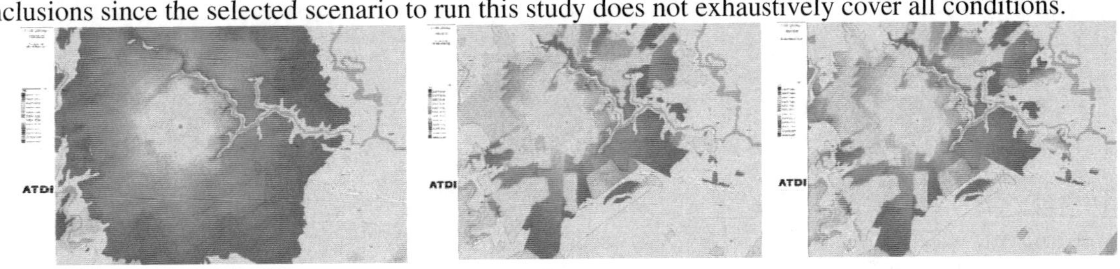

Scenario 1 - no clutter info

Scenario 2 - with clutter but no clutter attenuation

Scenario 3 - with fine-tuned clutter parameters

Figure 3. RF Coverage maps.

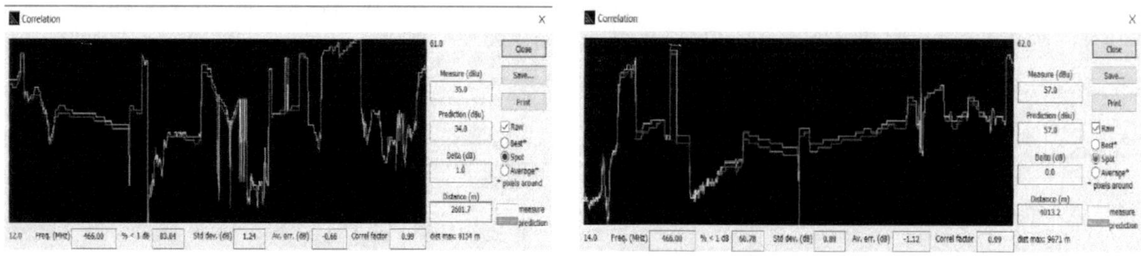

SD & AE Route 2 SD & AE Route 1

Figure 4. SD and AE in correlation results for Scenario 3.

Table 1. Coverage maps and correlation results.

Correlation results	Scenario 1	Scenario 2	Scenario 3
SD - Route 1	3.52 dB	3.32 dB	0.89 dB
AE - Route 1	-0.27 dB	0.88 dB	-1.12 dB
SD - Route 2	3.55 dB	1.66 dB	1.24 dB
AE - Route 2	0.76 dB	-0.75 dB	-0.66 dB

Table 2. Tuned clutter parameters.

Clutter type	open/ rural	village	suburban	urban	forest	parks	water	corn fields	common crops
Attenuation [dB/km]	1	6	8	12	6	5	0	4	3
Height [m]	0	4	8	16	25	8	0	2	1

5. References

[1] Grecu V, Bindar V, Halunga S, Popescu M 2015 *Comparative analysis of propagation models used in making coverage predictions for TETRA networks* Int. Symp. on Signals, Circuits and Systems) 9-10 July 2015
[2] Meng Y S, Lee Y H and Ng B C 2009 Progress in Electromagnetic Research **B17** 117
[3] Li L W, Lee C K, Yeo T Sand M. S. Leong, 2004, IEEE Trans. Antennas Propagation **52** 2445
[4] J. O. Famoriji, and Y. O. Olasoji, 2013 *Int. J. of Science and Technology* **2** 4 p164
[5] Weissberger M A 1981 *An initial critical summary of models for predicting the attenuation of radio waves by foliage* Electromagnetic Compatibility Analysis Center, Annapolis, USA
[6] Popoola S I, Atayero A A, Faruk N, Calafate C T, Adetiba E, Matthews V O 2017 *Proc. World Congress on Engineering* 5-7 July 2017 UK
[7] Pušnik T, Teković A, Zelenika Z 2015 *Proc. 25th Int. Conf. in Radioelektronika*, 21-22 April 2015
[8] Parsons J D 2000 *The Mobile Radio Propagation Channel* J. Wiley & Sons Ltd. pp. 15–31
[9] Almajanu F, Nastase C, Martian A, Marghescu I 2016 *U.P.B. Sci. Bull., Series C*, **78** 2, p 177.
[10] ITU-R P.526-13, *Propagation by diffraction* 2013, Series of ITU-R Recommendations
[11] Recommendation ITU-R P.676-3 1997 *Attenuation by atmospheric gases* ITU Radiocommunication Assembly
[12] Parmar K J, Nimavat V D 2015 *Int. J. of Innovative Research in Computer and Communication Engineering*, **3** 2 p840
[13] ITU Recommendation P.837 *Characteristics of precipitation for propagation modelling* https://www.itu.int/rec/R-REC-P.837/en, accessed October 2017
[14] Deygout 1966 IEEE Trans. on Antennas and Propagation **AP-14** 4, p 480
[15] Lee W C Y 1993 *Mobile Communications System Design* Wiley (Michigan) USA
[16] Militaru G, Popescu D, Ichim L, 2018, *UHF propagation model calibration using correlations between field measurements and prediction* IEEE Radio and Antenna Days of the Indian Ocean (RADIO2018). 1-2. 10.23919/RADIO.2018.8572437.

IOP Conf. Series: Materials Science and Engineering **524** (2019) 012004 doi:10.1088/1757-899X/524/1/012004

Electric field intensity in cross-section of evaporation duct

D G Serebriakov and R R Abdullin

Engineering School of Information Technologies, Telecommunications and Control Systems, Ural Federal University named after the first President of Russia B.N.Yeltsin, 19 Mira Street, Yekaterinburg, Russian Federation

r.r.abdullin@urfu.ru

Abstract. This paper is devoted to the investigation of an evaporation duct in the context of its impact on microwave links communication stability. The work contains the numerical results for previously developed mathematical model based on using the dyadic Green's functions. The electric field intensity dependencies on the receiving antenna suspension height and height of evaporation duct have been obtained for two variants of antenna polarization: horizontal and vertical. These results make it possible to formulate recommendations on how to place the transceiver antennas of microwave links to ensure 24-hour radio communication stability.

1. Introduction

The people, operating the microwave retransmission systems, periodically face the problem with the radio communication stability on long-distance intervals, especially spread over the water objects such as lakes and rivers. In this case, the line of sight presence might be insufficient for an acceptable signal level. A perfectly working daytime microwave link is able to malfunction in the morning.

The reason for this is highly likely to be a fog presence over the water surface. The fog density continuously decreases against the altitude, which can be described by the gradient of a moisture and, consequently, by the gradient of an effective permittivity. Therefore, when constructing the electrodynamic model it is possible to divide the above-water space into n separate layers with their own finite dielectric constant so as the gradient appears when the number of layers n tends to infinity. This representation makes it possible to clearly imagine why the connection is being lost. Traditionally, the overland propagation is described with a two-ray model. In the other words, the electric field intensity at the reception point is the result of simultaneous contribution of the direct ray and ray reflected from the underlying surface (from the water surface in the current case). However, this complex medium is nonhomogeneous and more optically dense in the bottom part. That is why, returning upward the radio waves reflected from the water surface, under certain conditions, undergo an effect of total internal reflection and turn out to be guided. Taking into account the very nature of the fog, the phenomenon described above was called as an evaporation duct.

If the transceiving antennas at the both ends are located inside the evaporation duct, the signal transmitted can be additionally enhanced in comparison with the free space propagation [1]. However, sometimes one of antennas (for example, transmitting one) is placed above the evaporation duct. In this case, getting in the duct, the wave, direct in daytime conditions, refracts and deviates from the initially right direction, thereby disrupting the connection between two distant ends.

In order to understand whether it is possible to improve the communication stability by means of relatively simple manipulations with transceiving antennas, we have suggested considering the evaporation duct as a multilayer structure in [2]–[3]. We have also adapted an apparatus of dyadic Green's functions for its analysis. In contrast to the methods of geometrical optics, Green's functions make it possible to take into account some crucial nuances such as the antenna pattern and multipath propagation without deep understanding a field interaction between the layers of model and without the necessity to know how many times the signal can reflect inside the duct until it reaches the receiver. We made some reasonable approximations, which then were validated with an example of Hertz dipole in a homogeneous medium. The present work is aimed to obtain numerical results for this model and further develop recommendations for the optimal antennas placement.

2. Model of Evaporation Duct

The developed model of the evaporation duct is presented in figure 1.

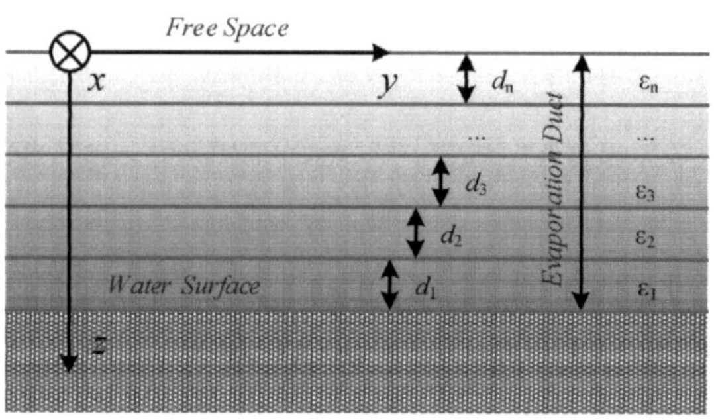

Figure 1. Model of the evaporation duct. This is the endless sandwich structure, layers of which are parallel to the water surface. Every layer has its own thickness d_p and permittivity ε_p. The coordinate axes are chosen as follows: z-axis is normal to the layers and directed downward; x and y lie in the plane parallel to the water surface and form the right system with z-axis; the origin coincides with the transmitting antenna position.

At the first stage of simulation we have taken as the transmitting antenna an electric dipole with uniform current distribution along its length. We considered two different cases of dipole orientation, which correspond to two orthogonal variants of linear polarization. The dipole placed along the x-axis is an equivalent of horizontally polarized antenna, whereas along-z-axis orientation means the vertical polarization. Therefore, in order to find the electric field intensity in arbitrary point of space, it is only necessary to calculate three components of Green's function per situation: Γ_{11xx}, Γ_{11yx}, Γ_{11zx} for horizontal polarization, and Γ_{11xz}, Γ_{11yz}, Γ_{11zz} for vertical one. However, if the receiving and transmitting antennas are perfectly aligned, the components Γ_{11yx}, Γ_{11zx} for the first case and Γ_{11xz}, Γ_{11yz} for the second case will be equal to zero at the receiver site. Consequently, the vector of electric field intensity is turned out to be completely specified by the component to which the dipole is parallel.

Γ_{11xx}-component is possible to be found from [2]. The calculation of Γ_{11zz}-component is occurred in accordance with much the same sequence of actions. Firstly, to simplify expressions, we introduced

$$\rho = k_\perp\left[(x-x')^2+(y-y')^2\right]^{1/2},$$

where $k_\perp$ is a transverse wavenumber defined as a geometric sum of k_x and k_y, which are the propagation constants along x and y axes, respectively. The dashes here mean the coordinates of transmitter; x' equals to zero for vertically positioned dipole, whereas y' has zero value in both cases considered. Then we transformed an inner integral in general formula for Γ_{11zz}-component from [4] to zero-order Bessel function of the first kind $J_0(\rho)$, and finally used the saddle-point method:

$$\Gamma_{11zz} = \sqrt{2\pi}\cdot f_{zz}(k_{\perp 0})\cdot\left[S''_{zzE}(k_{\perp 0})\right]^{-1/2}\cdot\exp\left[S_{zzE}(k_{\perp 0})\right], \quad (1)$$

where $f_{zz}(k_\perp) = J_0(\rho)\cdot k_\perp^3/(2\pi f\varepsilon_a)$, f is an operating frequency; $S_{zzE}(k_\perp) = \ln\left[\overleftrightarrow{Y}^E(0)/\gamma\overleftarrow{Y}^E(0)\right] - i\gamma z$, $\overleftarrow{Y}^E(0)$ is input E-plane modal admittance recalculated from the water surface to the origin by the recurrence formulas according to [4], $\overleftrightarrow{Y}^E(0) = \overleftarrow{Y}^E(0) + \overrightarrow{Y}^E(0)$, $\overrightarrow{Y}^E(0)$ is top-side input admittance, which equals to equivalent wave admittance of free space for guided TM-waves if transmitting antenna is placed above the evaporation duct; ε_a and γ are the absolute permittivity and longitudinal (along z-axis) propagation constant of the model layer closest to the point of observation; $k_{\perp 0}$ is the

IOP Conf. Series: Materials Science and Engineering **524** (2019) 012004 doi:10.1088/1757-899X/524/1/012004

root of the equation $\partial S_{zzE}(k_\perp)/\partial k_\perp = 0$. The use of equation (1) is possible independently on where the antennas are placed in relation to the evaporation duct. However, if the transmitting antenna is inside the duct, the second summand in expression for S_{zzE} gets significantly complicated, since the top-side input admittance is also to be recalculated by the recurrence formulas.

Thus, the analysis of equation (1) and all variables composing it allow concluding that in order to find the electric field intensity in arbitrary point of space, it is only necessary to correctly find the recalculated to the reference plane admittances, which are completely defined by the thicknesses and complex permittivities of layers into which the whole model is divided. However, it is obvious that resulting analytical expressions for field intensity are hardly possible to get even for two-layered structure due to their awkwardness. Therefore we used computer calculations.

3. Simulation Results

Rephrasing the words from previous section, we can conclude that, in order to determine the electromagnetic field intensity in the evaporation duct, it is only necessary to substitute the distribution of permittivity with height. In a simulation performed by means of MATLAB software we used the linear profile in the range of real numbers from 1.5 at the water surface to 1 at the upper bound of the duct, as it suggested in [5]. We have omitted the imaginary part of permittivity responsible for the losses in the media because it is extremely small at such super high frequencies as the traditional frequencies of retransmission systems are. For example, the dielectric loss tangent of distilled water approximately equals to 10^{-4}. Consequently, an attenuation constant for wave propagating inside it is very close to 0.08 dB/m. The loss tangent and attenuation in vapour depend on density (concentration) of the fog, and here they are significantly less. Thus, it can be neglected at the first stage of analysis.

We took $n = 100$ as the number of dielectric layers in the model, and $l = 0.01$ m as the dipole length so as it certainly fits in thickness of single slab for vertically oriented radiator. We considered the situation when the transmitting antenna was strictly above the duct – at an altitude of 25 meters, since in many models this value definitely exceeds the mean of evaporation duct height [6]–[7]. At the same time, the evaporation duct cannot appear and disappear immediately, which means that its height varies with time. Therefore, our task was to find "zones of stability" at a certain height. In the other words, it was necessary to detect such receiving antenna suspension heights at which the observed change in the electric field intensity is negligible regardless of the evaporation duct size.

Figures 2 and 3 show the intensity of horizontally and vertically polarized electric field at the receiver position vs. the receiving antenna suspension height and the height of the evaporation duct. The length of considered microwave link is 13.2 km. We believe that communication quality at the existing microwave links is acceptable during the day by default. That is why we decided to normalize the values obtained to the values of day propagation, i.e. to the case when it is possible to use the conventional two ray model. This way of data presentation lets intuitively understand how much worse or better the signal level is in the morning hours in comparison with an afternoon. Figures 2 and 3 make it clear that these "zones of stability" do exist in cross section of the evaporation duct.

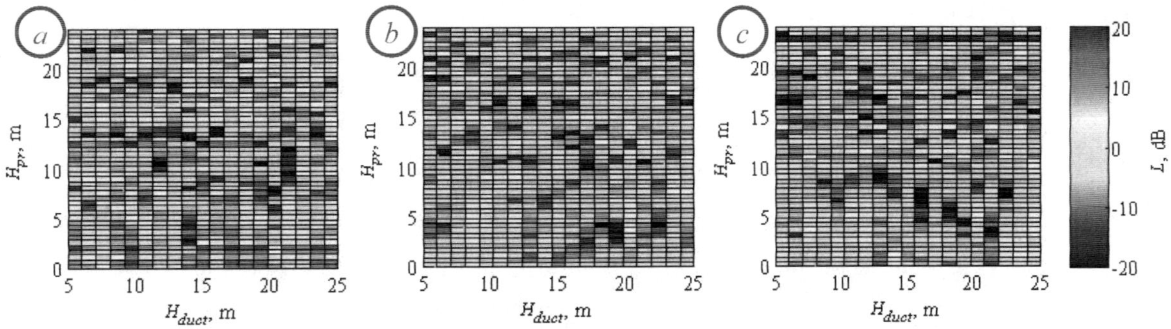

Figure 2. The relation between horizontally polarized electric field intensities at the receiver and transmitter sites, normalized to the same in free space: (*a*) 8 GHz, (*b*) 13 GHz, (*c*) 18 GHz.

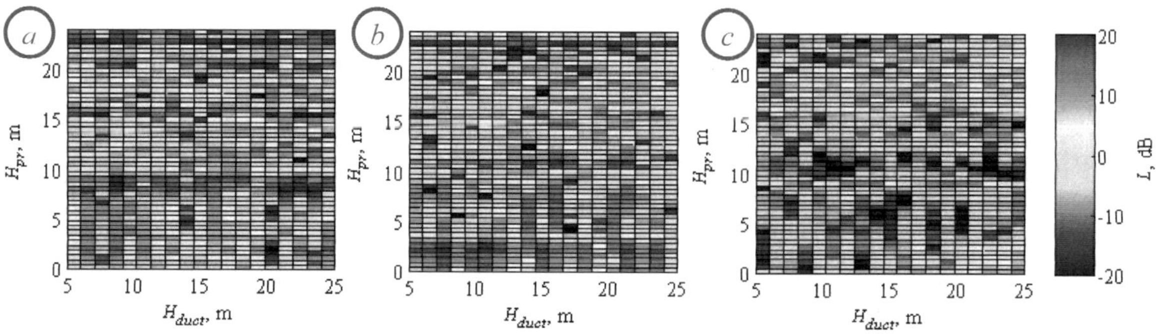

Figure 3. The relation between vertically polarized electric field intensities at the receiver and transmitter sites, normalized to the same in free space: (*a*) 8 GHz, (*b*) 13 GHz, (*c*) 18 GHz.

For instance, the optimal suspension height of horizontally polarized receiving antenna is 12.5 m at 8 GHz, 18 m at 13 GHz, and 14.5 or 23.5 m at 18 GHz. At the same time, the vertically polarized antennas have a generally better reception under conditions of evaporation duct, even in comparison with the daytime propagation, probably due to the possible occurrence of pure refraction. Here the best values of suspension height are 23.5 m at 8 and 13 GHz, and 9 m at 18 GHz.

4. Conclusion

The present paper provides the numerical results of evaporation duct digital simulation. The analysis of obtained two-dimensional dependencies made it possible to obtain the optimal position of both horizontally and vertically polarized antennas in order to ensure the minimal fluctuations of electric field intensity at the receiver site. The study undertaken for three standard microwave link frequency bands has shown that in spite of difference in particular situations the use of vertical polarization is generally preferable. The results of research allow concluding that the correct adjustment of the receiving antenna suspension height does reduce the influence of evaporation duct on communication quality at the microwave links. Nevertheless, rigorous formulation of recommendations requires more experiments to be conducted.

Acknowledgments

The work was supported by Act 211 Government of the Russian Federation, contract № 02.A03.21.0006.

The research was executed by the grant of the Ministry of education and science of the Russian Federation (project № 8.2538.2017/4.6).

References

[1] Levadnyi Yu V, Ivanov V K and Shalyapin V N 2010 About relationship of evaporation duct height and microwave attenuation in over horizon *Radiophysics and Electronics* **1** no 4 pp 45–50.

[2] Serebriakov D G and Abdullin R R 2018 Model of evaporation duct. *Proc. Ural symposium on biomedical engineering, radioelectronics and information technology (Yekaterinburg)*.

[3] Serebriakov D G and Abdullin R R 2018 Optimal Antenna Position Inside Evaporation Duct. *Proc. IEEE Radio and Antenna Days of the Indian Ocean (RADIO)*.

[4] Panchenko B A, Nefedov E I 1986 Microstrip antennas. (Moscow: Radio & Connection).

[5] Paulus R A 1990 Evaporation duct effects on sea cluster *IEEE transactions on antennas and propagation* **38** no 11 pp. 1765-71.

[6] Devereux W E and Van Egmond J 1989 Effects of ducting on radar operation in the Persian Gulf *Proc. IEEE National Radar Conference (Dallas)* pp 174–181.

[7] Hitney H V and Hitney L R 1990 Frequency diversity effects of evaporation duct propagation *IEEE Transactions on antennas and propagation* **38** no 10 pp 1694–1700.

IOP Conf. Series: Materials Science and Engineering **524** (2019) 012005 doi:10.1088/1757-899X/524/1/012005

Novel Low THz Frequency Extenders for Vector Network Analyzers

D R Vizard[1], E G Hoare[2], M Gashinova[2] and M Cherniakov[2]

[1] VivaTech SARL, Nice, France
[2] EECE, University of Birmingham, Birmingham, UK

Email: drv@vivatech.biz

Abstract. The Low THz spectrum has potential for high resolution imaging applications such as automotive radars and high data rate communications. Propagation studies and test platforms require compact portable equipment which may be achieved by extending the frequency range of commercial battery powered vector network analyzers (VNA). The development of both desktop and portable 110 – 670 GHz VNA based extenders is described along with some recent experimental test data using these instruments.

1. Introduction

Autonomous navigation and 'hands-off' driving are widely anticipated features of future vehicles and mobile robots, where smart sensors must establish vehicle "cognition" under all weather conditions. The TeraHertz (THz) portion of the electromagnetic spectrum may be exploited to enable the high definition imaging radar systems required. Further, the needs of high data rate communications require the development of THz transmitter and receiver systems. Current solid state semiconductor THz technologies allow the implementation of VNA (Vector Network Analyzer) extenders and test instruments essential to the development of components and systems. Compact and battery portable equipment ideal for field testing can also be based on low cost commercial VNA's by adding appropriate THz extenders. This paper is an extended version of a presentation at the IEEE 2018 Radio Days, Maurititus, Oct. 2018. [4]

2. Desktop VNA Extenders

Laboratory VNA's are in extensive use but are in general limited to < 40 GHz coverage. Beyond this frequency 'add-on' extenders can be used to reach up to 1.1 THz, depending on the instrument in question. They are, however, bulky and best suited to laboratory environments. Such extenders rely upon the access to the internal oscillators and IF chain of the instrument. A common approach is to generate the higher frequencies by multiplication of the VNA internal microwave oscillators and use a harmonic mixer down-conversion process in the receiver. By using internal high-directivity couplers the full range of S-parameters can be extracted by comparisons of the transmitted and received power levels. An example of a 110 – 170 GHz VNA desktop extender is shown in figure 1. The conventional interface to a standard microwave VNA is described in figure 2.

The device under test (DUT) is placed between the waveguide ports of the two TR (transceiver) extender units. In this way, the S-parameters can be determined by exciting the DUT in transmission or reflection from either direction.

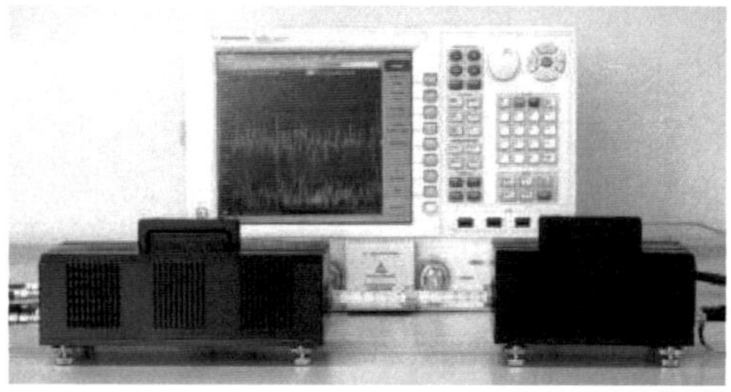

Figure 1. Low THz VNA Extender System

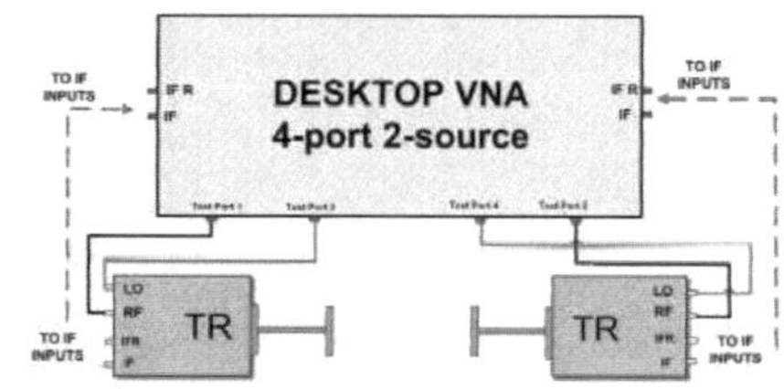

Figure 2. 110-170 GHz VNA with TR-TR Extenders

3. Portable VNA Extenders

Recent developments in compact portable battery powered VNA instruments allow their use in field applications such as tower testing in communications, spectrum monitoring and others. Such instruments are available up to 50 GHz and may be readily adapted to THz applications such as field trials of automotive radars and others.

Novel architectures are required for such extenders, however, because these portable instruments have no access to their internal microwave local oscillators and receiver IF chains. A solution that we have implemented uses up and down converters phase locked to the instrument 10 MHz reference clock to generate the required THz transmission and receiver functions. The dynamic range equals or exceeds that of conventional desktop instruments and enables their use in field trials of communication systems or automotive research. Further, interfaces are considerably simplified. An example of an extender configuration operating at 300 GHz is described in figures 3 & 4. The system forms the front end for an experimental automotive radar. The transmit power was -10 dBm (100uW) and the receiver noise figure is 14 dB (DSB).

Figure 3. Portable 300 GHz VNA Radar Front

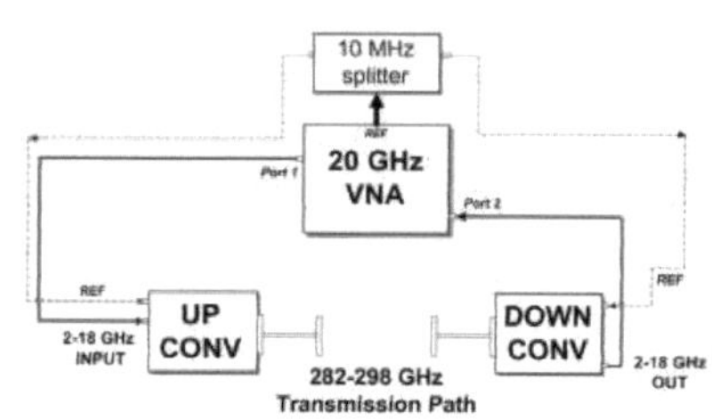

Figure 4. 300 GHz Front End Configuration

4. Radome Transmissivity Measurements

The 300 GHz VNA radar front end was used to measure the transmission properties of thin water films on radome material, and compare these with data and measurements at lower frequencies. Additionally, the effect of a range of contaminants found in typical automotive environments, such as salt, ice, sand, diesel and gasoline fuel was measured. Compilations shown in figures 5 (a) and (b) demonstrates that there is no dramatic difference between the transmissivity of thin water films at common automotive frequencies (24 and 77 GHz) and at low THz frequencies (150 and 300 GHz). Full details of this work are available in Ref [3].

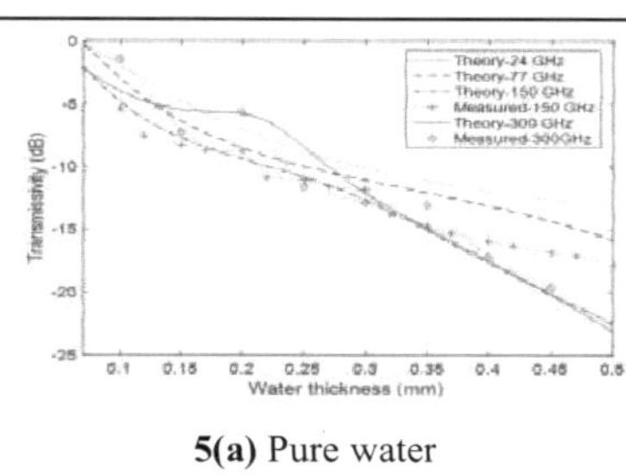

5(a) Pure water

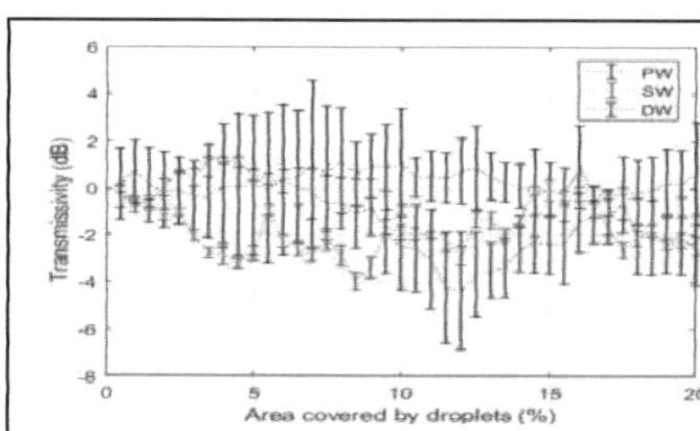

5(b) P Pure, S salt, D dirty

Figure 5. Comparative Transmissivity of Water Films 24-300 GHz

IOP Conf. Series: Materials Science and Engineering **524** (2019) 012005 doi:10.1088/1757-899X/524/1/012005

5. Radar Resolution Comparison Measurements 77/300 GHz

In order to demonstrate the improved resolution that may be achieved with the 300 GHz radar, a comparison was made between the low THz mobile test platform and a 77 – 79 GHz radar of equivalent antenna beam pattern. The radars were mechanically scanned over the same azimuth plane, viewing a typical mixed road scene (figures 6a, 6b, 6c).

Figure 6(a) Road Visual Scene

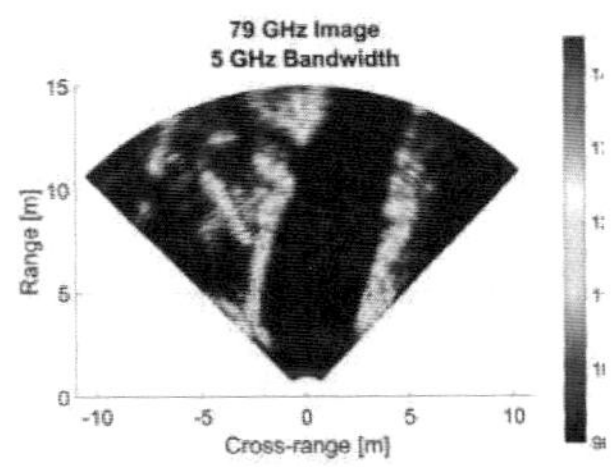

Figure 6(b) 79 GHz Image

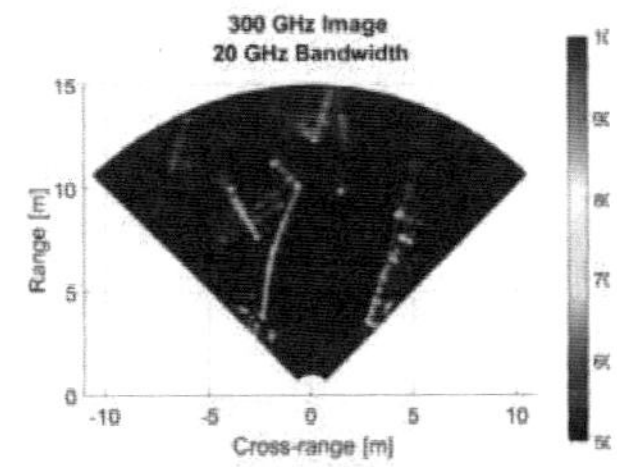

Figure 6(c) 300 GHz Image

6. A 656-672 GHz Hybrid VNA Extender

In order to further improve resolution and explore the THz transmission properties of the atmosphere and materials under adverse conditions (snow, ice, fog, dirt and dust) a 670 GHz extender was developed. A radar using this extender will be used to evaluate the achievable range and resolution when operating in this frequency band. Given the shorter wavelength obtaining sufficient transmitter power for an effective radar is challenging. The use of a converter approach as deployed at 300 GHz would provide inadequate output power due to the limitations of mixer up-converters at 670 GHz. This is due to the reduced output power of a mixer based up-converter which is constrained by : (a) saturation of the mixer, (b) available local oscillator drive power and (c) mixer conversion loss.

A comparison of these parameters for 300 GHz and 670 GHz mixer up-converters predicted an expected output power reduction of 10 – 15 dB for the higher frequency, (see table 1).

Table 1. Comparison of Mixer Upconverter Output Power at 300 and 670 GHz

Frequency (GHz)	Measured Output Power (dBm)	Extrapolated Output Power (dBm)
300	-15	
670		-25 to -30

Considering the 670 GHz output power estimates, a novel hybrid design was developed which provided up to 15 dB more power than an up-converter approach, by driving a multiplier chain from the converter/divider arrangement described schematically in figure 8. The method uses a X36 multiplier chain as the transmitter whilst maintaining the same frequency plan required by the portable VNA architecture. This was achieved by feeding the multiplier chain with a frequency divided by the same multiple, in this case 36, whilst applying a suitable frequency offset from a stable secondary source, phase locked to the VNA reference clock.

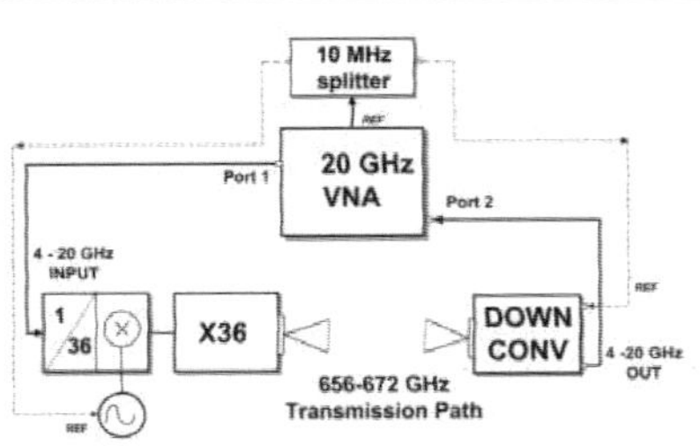

Figure 7. 656-672 GHz Hybrid VNA Extender

Figure 8. 670 GHz Extender with VNA

In this way an additional 15 dB of transmitter output power was obtained when compared with a mixer up-converter approach. The system produces - 10.5 dBm typical transmit power and has a receiver noise figure of 12 to 14 dB DSB in the band 656 – 672 GHz. These results represent state-of-art performance for wideband transmitter/receiver systems at this wavelength.

7. 670 GHz Water Film Transmission Measurement

The hybrid VNA extender was used to measure 670 GHz transmission losses in thin water films. A theoretical model based on Fresnel theory of reflection and transmission for multilayer structures, air-water-air, was adopted. The complex refractive index of a material $\tilde{n}(f,T) = n - jn$, which depends on both frequency and temperature, will result in reflection of some of the energy at the first medium interface and the rest of the energy will be transmitted through the second medium. The reflection and transmission coefficients of each layer depend on the permittivity, thickness, and shape of the layer. A water film of known thickness may be modelled from its conductivity, magnetic permeability and dielectric permittivity. The experimental test set-up is shown in figure 10, and a comparison between a theoretical model and measurements is shown in figure 11. Results confirm current models provide good estimates of 670 GHz transmission losses for water films in the thickness range 0.1-0.45 mm.

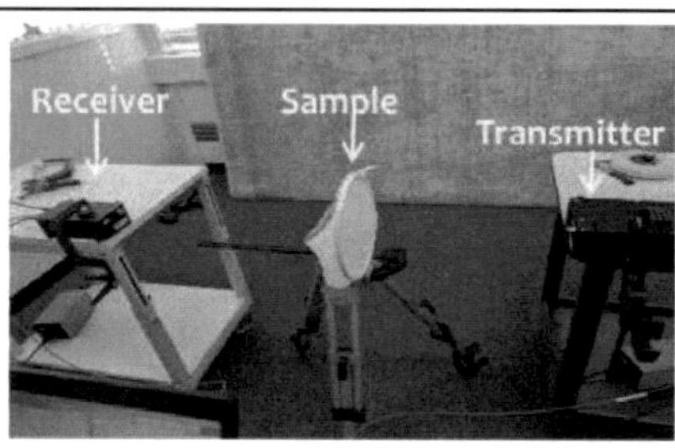

Figure 9. Transmission Test Arrangement

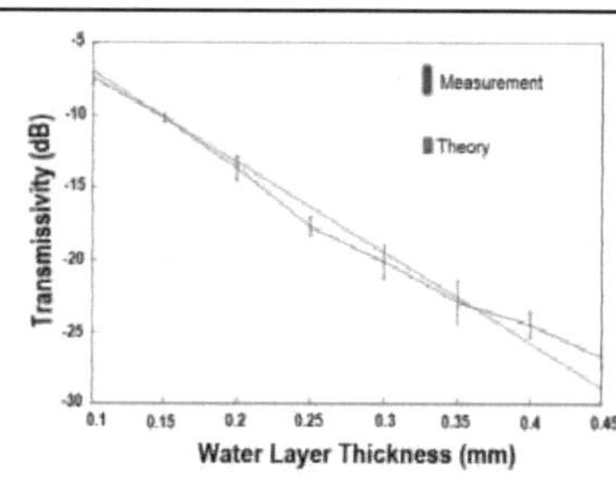

Figure 10. Water Thickness vs. Transmission

8. Conclusion

Following the success of 300 GHz propagation and radar imaging experiments, a 670 GHz hybrid VNA extender has been completed and will be used to (i) make atmospheric transmission measurements under adverse conditions and (ii) test radar performance. The development of this portable 670 GHz radar front end was described along with transmission tests of thin water films.

Acknowledgments

This work has been supported by Jaguar Land Rover Research Department, Coventry, UK.

9. References

[1] Appleby, R. and Anderton, R.N.,"Millimeter-Wave and Submillimeter-Wave Imaging for Security and Surveillance, *Proc. of the IEEE*, **V. 95**, N. 8, 2007

[2] D R Vizard et al, 'Antenna Range Evaluations of Low THz Imagers for Automotive Applications', *Intl. Conf. Antenna Measurements and Applications*. Juan-LesPins, France **APP3.2** Nov 2014

[3] F.Norouzian et al, 'Low THz Transmission Through Water-Containing Contaminants on Antenna Radome', *IEEE Trans. Terahertz Science and Technology*, vol.8, no1, pp.63-75, Jan 2018

[4] D R Vizard et al, 'Applications of Low THz Frequency Extenders for Vector Network Analyzers', *IEEE Radio and Antenna Days*, **P**.529, Wollmar, Mauritius, 16 Oct 2018

IOP Conf. Series: Materials Science and Engineering **524** (2019) 012006 doi:10.1088/1757-899X/524/1/012006

Microwave-Photonics Frequency Up-Converter for Telecom and Measurement Equipment

D A Fofanov, T N Bakhvalova, A V Alyoshin, M E Belkin and A S Sigov

Scientific and Technological Center "Integrated Microwave Photonics", Russian Technological University (MIREA), Moscow, Russian Federation
E-mail: belkin@mirea.ru

Abstract. An ultra-wideband tunable frequency converter scheme for microwave and millimeter-wave signal generation based on optical recirculating loop and microwave-photonics technique, is proposed and demonstrated. The scheme consists of standard, commercially available elements such as dual parallel Mach-Zehnder modulator, optical filter, wideband photodiode and does not require ultra-broadband RF components. A recirculating loop acts as an optical frequency comb generator and allows obtaining multiple local oscillator signals at frequencies that significantly higher than the RF reference used. Up-converted signals of frequencies in mm-wave band are demonstrated experimentally with spectra free of spurious products. The converter proposed could be used as an alternative solution to microwave mixers in telecom and measurement equipment.

1. Introduction

Recent advances in communication and radar technologies drive the growth of demand for test and measurement equipment [1]. As the frequencies overcome Ka-band threshold, the problem of ultra-wideband signal processing becomes more and more challenging [2] and modern radio-frequency (RF) measuring apparatuses are no longer able to meet the requirements [3]. Besides, utilizing license-free mm-wave bands is a possible solution for broadband communication in next generation wireless networks [4]. For example, wide bandwidth available between 60 and 90 GHz enables wireless links with more than 10-Gbps throughput that is significantly higher than at lower RF bands.

Communications at such high frequencies pose unique challenges, one of which is an efficient frequency conversion. A common approach to up- and down-conversion in communication hardware as well as in a measurement equipment relies upon electronic mixers [5]. As the frequencies rise to a sub-THz domain and bandwidth exceeds octave, these mixers become more and more unpractical. Traditional mm-wave frequency conversion networks tend to be narrowband, bulky and costly that prompts us to look for alternative solutions.

A novel microwave photonics (MWP) approach based on electronic-to-optical conversion of RF signals and their optical domain processing has established itself as a promising solution [6]. The basic principle of photonic signal processing is to convert RF signal into optical domain, for example via intensity modulation of light, carry out some operations in optical band, and then demodulate it back into RF. The main idea is that some types of processing are easier to perform in optical domain. This paper presents a novel design concept and key experimental performances of MWP up-converter for telecom and measurement equipment that is proposed for overcoming the above-mention challenges of mm-wave applications.

2. Principle of Operation

Figure 1 shows the schematic diagram of proposed wideband MWP frequency up-converter. The lightwave is emitted from a continuous wave (CW) narrow-linewidth semiconductor laser and being split into two branches by optical coupler (OC). The upper branch contains a dual parallel Mach-Zehnder modulator (DPMZM) for single-sideband suppressed-carrier (SSB-SC) electro-optic modulation (EOM) of lightwave with intermediate frequency (IF) signal. This DPMZM is biased properly to suppress lower sideband of optical spectrum; its output optical field *E(t)* is described in simplified form by

$$E(t)=E_0\cos[2\pi(\upsilon_0+f_{IF})t], \tag{1}$$

where E_0 is an optical field amplitude, υ_0 is a laser frequency and f_{IF} represents IF signal frequency.

DC bias is applied to the DPMZM using so-called modulator bias controller (MBC). It's primary function is to track for modulator's transfer function drift and to move modulator's operating point (bias) in a coordinated manner. Tracking is accomplished by applying low-frequency (kHz-range) pilot tone to the DPMZM and processing response of the monitoring photodiode that is embedded in the modulator.

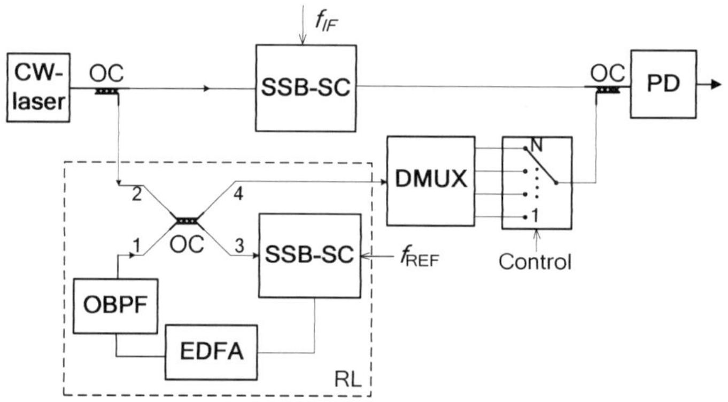

Figure 1. Schematic diagram of the ultra-wideband frequency up-converter.

The lower branch is intended to form a tunable local oscillator (LO) signal and contains an optical recirculating loop (ORL), a wavelength division demultiplexer (DMUX) and N-position electrically controllable optical switch element.

The ORL acts as an optical frequency comb generator. It consists of another SSB-SC modulator, an erbium doped fiber amplifier (EDFA) to compensate for losses in the loop and an optical bandpass filter (OBPF) for optical spectrum conditioning. As the laser signal passes the loop round by round, its spectrum takes the form of a frequency comb with a spacing, equal to the reference frequency f_{REF}. ORL operates as follows. DPMZM performs EOM of the reference signal f_{REF} from the RF synthesizer. This modulator is biased to suppress upper sideband of optical spectrum and its output is mathematically represented by equation (1) with a change of sign between υ_0 and f_{IF}. After amplification in EDFA, the signal passes through an OBPF that rejects the spectral components outside the operating frequency band.

Note that the comb period can be easily fine-tuned by changing the synthesizer reference frequency. The total number of usable LO frequencies N is given by

$$N=\Delta f_{PD}/\Delta f_{IF}, \tag{2}$$

where Δf_{PD} is a PD bandwidth and Δf_{IF} is the expected bandwidth of IF signal.

The output radio frequency (RF) signal is produced as a result of beating an upper sideband (IF) and a lower sideband (LO) optical signal components at the photodetector (PD). The lower sideband component is selected by simply switching to the proper spectral demultiplexer output that corresponds to the desirable operating RF band. A stable performance over the entire operational range is provided by the flat frequency response of PD.

IOP Conf. Series: Materials Science and Engineering **524** (2019) 012006 doi:10.1088/1757-899X/524/1/012006

This circuit is also capable to down-conversion. RF and IF change their places and one of the modulators is reconfigured to provide lower sideband modulation instead of upper sideband. This is done by simply changing a DC control voltage applied to the modulator by MBC, so it is reconfigurable indeed.

SSB modulator in a fiber loop is not the only way to generate frequency comb. Circuits with mode-locked lasers and Fabry-Perot modulators are known [7, 8]. They require complex stabilization mechanisms whereas the circuit proposed produces accurate frequency comb without stabilization because the frequency shift is completely determined by accuracy of RF reference signal.

3. Experimental Results

An experimental investigation of the proposed scheme was implemented following the setup of figure 1. It includes Teraxion PS-LM distributed feedback laser module as a source of light with a frequency of υ_0=193.317 THz, iXblue MXIQ-LN-40 DPMZMs in both branches controlled by iXblue MBC-IQ-DG-BT MBCs, Ericsson EDFA, optical switch and couplers from Opneti, and Finisar XPDV3120R 70-GHz photodiode for optical-to-electrical conversion and mixing. To date, commercially available DMUXs are designed for frequency steps of 50 GHz and 100 GHz. So, they are not able to select a separate comb component with a lower step. For this reason, the DMUX in our setup was constructed by the corresponding set of Teraxion's tunable OBPFs with less than 1-GHz bandwidth.

Figure 2 shows the upper-branch output representing the SSB-SC-modulated IF signal for two IF frequencies using in the experiment. The carrier suppression ratio is more than 30 dB and the lower sideband suppression ratio is about 20 dB. This parameter is determined by the extinction ratio, the branch balance of the modulator and by accuracy of the DC control voltage setting.

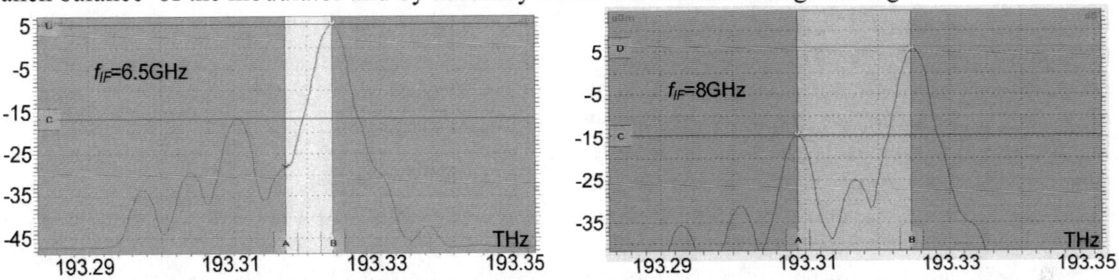

Figure 2. Output of the upper branch. f_{IF} = 6.5 GHz (left), f_{IF} = 8 GHz (right)

Optical frequency comb spectrum at the output of the ORL is shown in figure 3. Comb period is determined by reference frequency, useful LO frequencies are determined by equation (2) and amplitude equality depends on response flatness of the EDFA and the OBPF. Figure 4 exemplifies the OBPF's output centered at the frequency of $\upsilon_0 + 34$ GHz (reference doubled). The adjacent optical channels are suppressed by approximately 40 dB. This signal will ultimately beat at the PD to produce an up-converted output, so side channel suppression is important for spectrum purity of the resulting RF signal.

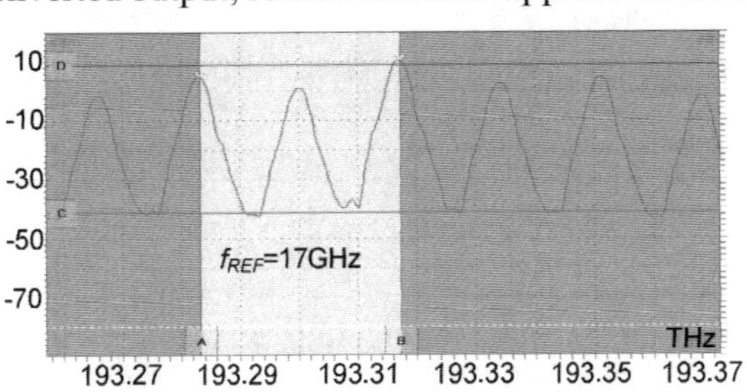

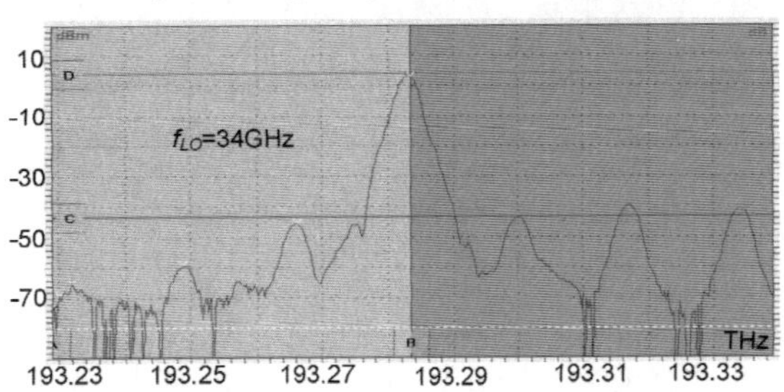

Figure 3. Output of the optical recirculating loop, f_{REF} = 17 GHz.

Figure 4. Output of the OBPF, f_{LO} = 34 GHz.

An example of RF spectrum of the PD output measured by Keysight E4448A spectrum analyzer is illustrated in figure 5. The output power falls from -19.5 dBm at 6.5 GHz (a) and 8 GHz (b) to -30 dBm at 40.5 GHz (c) and 42 GHz (d). Conversion loss is about 40 dB in reference doubling mode. This low signal level can be explained by low sensitivity of the applied PD. There is a compromise between PD bandwidth and sensitivity. High bandwidth leads to low sensitivity. Nevertheless, RF amplification is mandatory in any case.

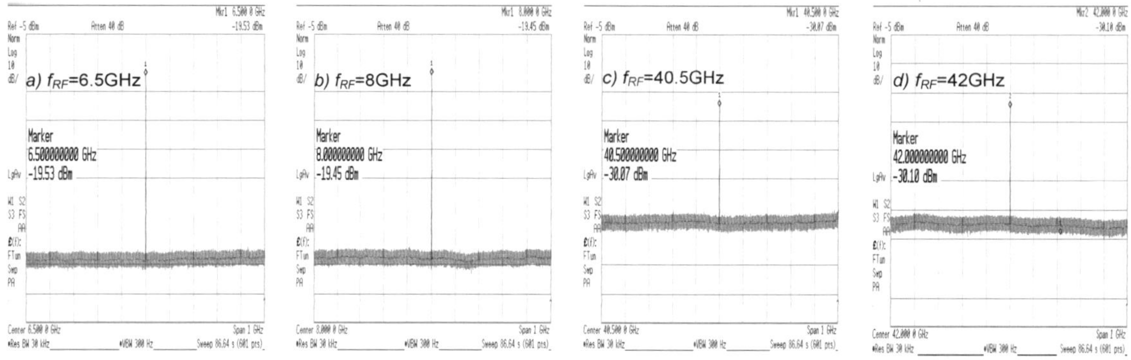

Figure 5. RF spectra of the PD output: a) f_{RF}=6.5 GHz; b) f_{RF}=8 GHz; c) f_{RF}=40.5 GHz; d) f_{RF}=42 GHz.

What is important is that the spectrum is clear and contains no spurious products, so this RF output has the potential of practical use. This experiment validates the principal applicability of the proposed scheme in communication hardware as well as in a measurement equipment when using an appropriate amplifier.

4. Conclusion

A microwave-photonics-based frequency up-converter scheme is proposed for mm-wave signal generation. This converter is of relatively simple design, does not require high frequency RF components and relies upon off-the-shelf commercially available elements. A high quality C-to-V band up-conversion is realized experimentally through all-optical processing with as low as 17-GHz reference. During these experiments, we were limited by the photodiode available in our disposal, but the potential of the scheme itself is bounded only by photodiode bandwidth, and there are already more than 100 GHz off-the-shelf devices. Thereby photonic frequency conversion is a promising approach in mm-wave band, despite the relative complexity and technology immaturity. Therefore, we are concentrated today at applying the proposed scheme for communication link in unlicensed 60-GHz band studying its key characteristics and quality metrics.

Acknowledgements

This work was supported by the Russian Foundation for Basic Research, Grant No. 18-58-80041.

References

[1] Pan S L and Yao J P 2017 Photonics-based broadband microwave measurement *IEEE/OSA Journal of Lightwave Technology* **35** no 16 pp 3498–513.

[2] Ng W 2015 Photonics for microwave systems and ultra-wideband signal processing *Optics Communications* **373** pp 2–15.

[3] Zou X, Lu B, Pan W, Yan L, Stöhr A, Yao J 2016 Photonics for microwave measurements *Laser Photonics Rev.* **10** no 5, pp 711–34.

[4] Browne J 2018 What Role Will Millimeter Waves Play in 5G Wireless Systems? *Microwaves & RF* **57** no 4 pp 38–42.

[5] Maas S 2003 Nonlinear Microwave and RF Circuits (Norwood: Artech House).

[6] Urick V J, McKinney J D, and Williams K J 2015 Fundamentals of microwave photonics (Hoboken: John Wiley & Sons).

[7] Diddams S A 2010 The evolving optical frequency comb *J. Opt. Soc. Am.* B **27** no 11, pp 51–62.

[8] Kim J, Richardson D J and Slavík R 2016 Cavity effect on phase noise of Fabry-Perot modulator-based optical frequency comb *IEEE Photonics Conference (IPC)* pp 503–4.

IOP Conf. Series: Materials Science and Engineering **524** (2019) 012007 doi:10.1088/1757-899X/524/1/012007

Compact directional coupler with five stubs

Denis A. Letavin

Ural Federal University named after the first President of Russia B.N. Yeltsin, Yekaterinburg, Russian Federation.
d.a.letavin@urfu.ru

Abstract. In this paper, a compact design of a five-stub microstrip 3-dB coupler with significantly reduced overall dimensions has been developed and investigated. To downsize the conventional layout, artificial transmission line segments, whose characteristics are equivalent to quarter-wave segments at the frequency of 2000 MHz and its neighbourhood, were used. This enables to obtain an area of the compact coupler which is 83.15% less than the conventional one, with just 26.15 % reduction in the operating frequency band.

1. Introduction
A branch-line (or 3-dB) coupler is a four-port device designed to divide the power incoming from one port equally between two other ones, whereas the fourth port remains isolated. An interest in miniaturizing the devices of this type is primarily contingent on the tendency to introduce microwave devices in the composition of integrated circuits while maintaining their performance. The design of the five-stub directional coupler is a modified design of a two branch-line coupler, which includes additional quarter-wave stubs in order to increase the operating frequency band. The coupler considered can be used for the division and the summation of power, as well as for feeding antenna systems. Since the conventional five-stub coupler has a quarter-wave width and, a length equal to the wavelength of the transmission line, the lower the operating frequency, the larger the area occupied. Today, there are a large number of possible methods to effectively reduce the dimensions of microstrip devices [1-5]. This work is aimed at miniaturizing a four-port coupler using original microstrip structures, equivalent to quarter-wave segments in a given frequency band [6].

2. Design coupler
The microstrip branch-line couplers used around the world are based on different design solutions. However, the difference generally consists just in the wave impedance values of the quarter-wave segments. The purpose of the present work is to miniaturize the five-stub coupler design so that the characteristics of the device are deteriorated minimally. The development and investigation of the designs are carried out using the NI AWR Design Environment software. The commonly used 1-mm-thick FR-4 substrate with relative permittivity ε_r = 4.4 and loss tangent tanδ = 0.02 was used as material. The block diagram and layout of the conventional design with the calculated impedances are shown in figures 1 and 2, respectively. The results of S-parameters and phase differences are shown in figures 3 and 4, respectively. The occupied area turns out to be 31.6 mm x 80.6 mm = 2547 mm^2.

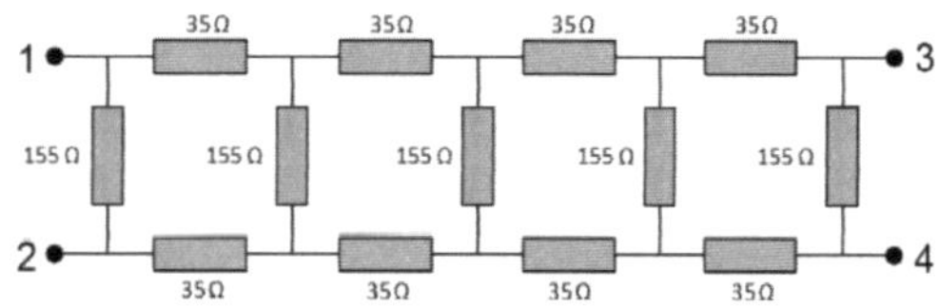

Figure 1. Block-diagram of a five-stub coupler tuned to the frequency of 2000 MHz.

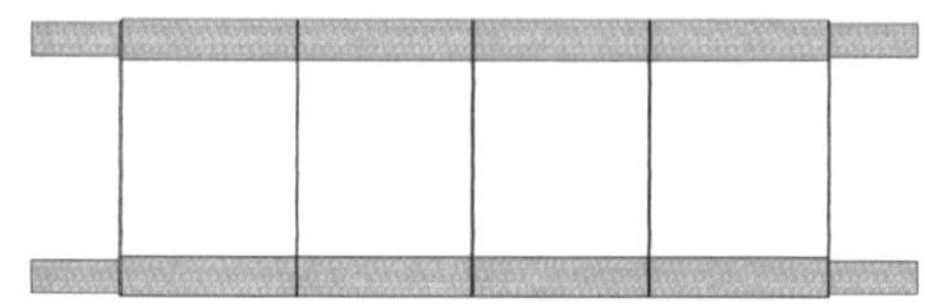

Figure 2. Conventional layout of a five-stub coupler tuned to the frequency of 2000 MHz.

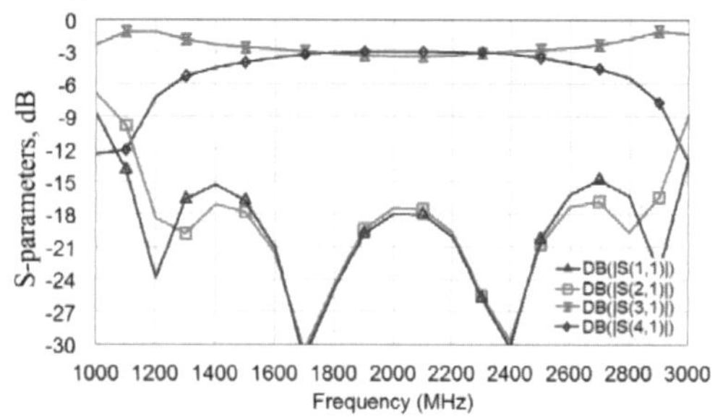

Figure 3. S-parameters of a traditional coupler based on Fig. 1 designed at a centre frequency of 2000 MHz.

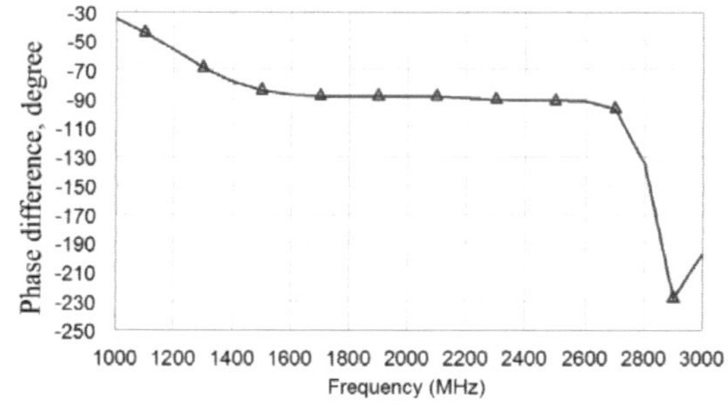

Figure 4. Phase difference between the signals at the coupler outputs for a coupler tuned to the frequency of 2000 MHz.

Based on the calculated frequency characteristics, it is possible to determine that the central frequency of operation is 2000 MHz, and the corresponding phase difference reaches 90 degrees. The frequency band estimated at a decoupling level of –15 dB amounts to 1758 MHz, or 87.9% bandwidth. The S11 parameter does not exceed –15 dB within this whole frequency band. The next step is the development of the electrodynamic structures with a phase shift of 90 degrees at the central frequency. These structures are shorter but wider than the conventional microstrip segments. However, the protruding elements of the newly developed structures are fitted in the free space inside the coupler.

The synthesis of the electrodynamic structures consists in the following steps:

1) Select the resistance value of high-impedance lines (which play the role of inductances), for example 100 Ohm, as well as their length.

2) Determine the capacitance value using the resistance formula. Knowing the values of the inductance and the resistance, it is straightforward to deduce the nominal value of the required capacitance.

3) Recalculate the values of the elements from the equivalent scheme to the parameters of the corresponding microstrip layout.

The simplest way to implement the inductance is to use a long segment of the transmission line. The implementation of the capacitance can be different. It is a low-impedance segment, or open circuit stub, whose length is less than a quarter wavelength. In the present work, we realized the capacitance as an open circuit stub bent in the form of a meander. The gaps between adjacent elements are selected on the basis of the layout technological feasibility. Using the structures suggested we managed to reduce the dimensions of the 3-dB coupler. The electrodynamic structures have the same phase at the central frequency of operation as well in its neighborhood. This makes it possible to obtain the coupler compact design which has characteristics consistent with the performance of the full-sized device. The final version of the compact five-stub coupler is shown in figure 5. The area resulting from optimization turned out to be 13.9 mm x 30.9 mm = 429.5 mm^2. Thus, the area of the compact design is 83.15% less than the area of the conventional coupler. Since it is required to obtain the maximum reduction in dimensions, the number and sizes are selected in such a way as to maximally utilize the space inside the coupler. Because artificial transmission lines are used for the miniaturization, couplers act as low-pass filters, and they provide harmonic suppression in a certain frequency band. The results of S-parameters and phase differences are shown in figures 6 and 7, respectively. The electrical lengths of the conventional and artificial microstrip line segments are compared in figure 8.

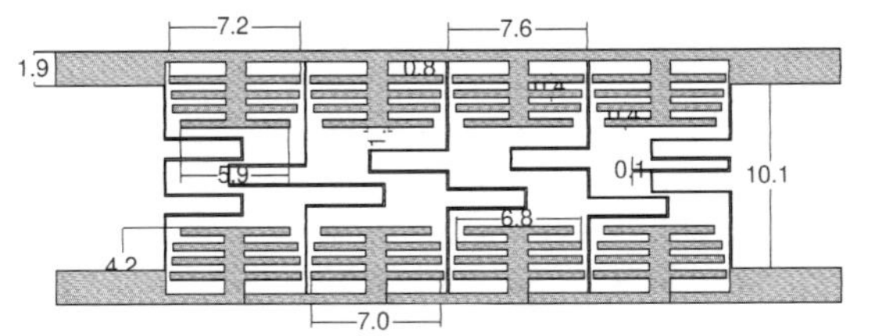

Figure 5. Layout of a compact 3-dB coupler tuned to the frequency of 2000 MHz.

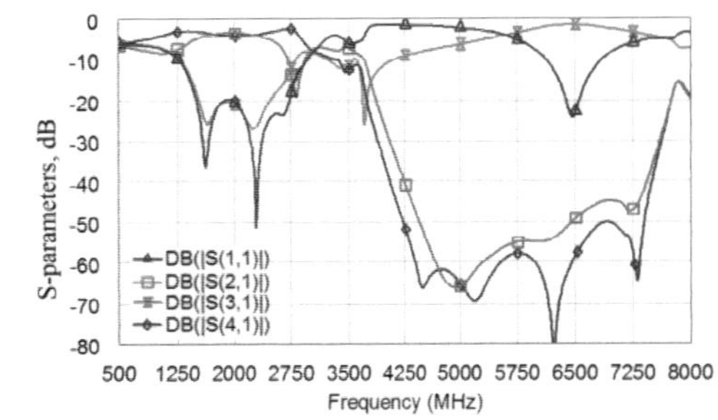

Figure 6. S-parameters of a compact coupler tuned to a centre frequency of 2000 MHz.

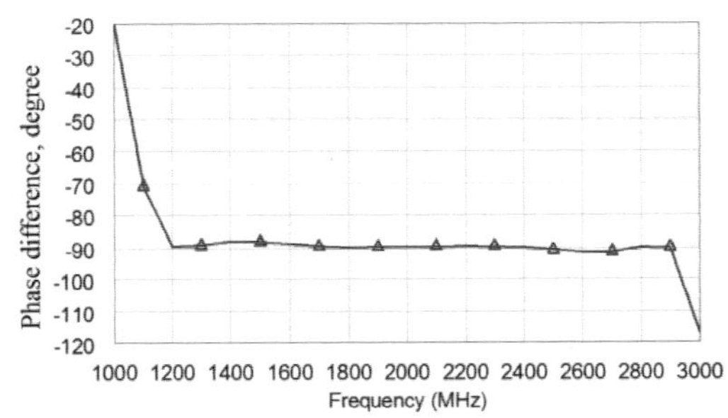

Figure 7. Phase difference between signals at the outputs of the coupler developed.

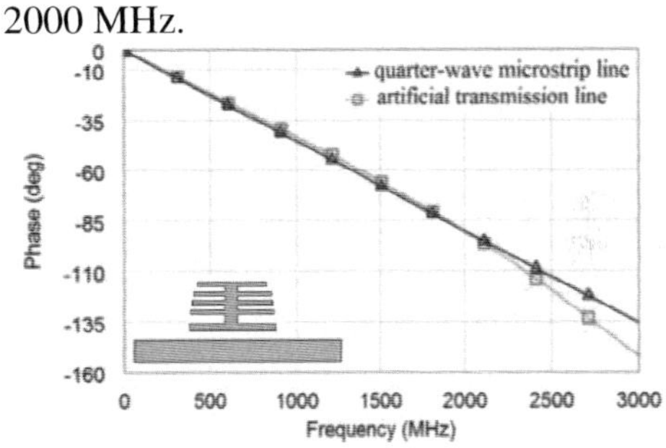

Figure 8. Electric lengths of conventional and artificial microstrip line segments.

For further analysis, it is necessary to verify the results of the numerical simulation with the measurement. For this purpose, using a method of etching printed circuit board we manufactured the 3-dB coupler prototype, shown in figure 9. After mounting the microwave connectors, we investigated this prototype by means of the vector network analyzer R&S ZVA24 operating in the frequency range from 10 MHz to 24 GHz. The measured S-parameters are shown in figure 10. The results of the numerical and experimental investigations are summarized in Table 1. It can be seen that the miniaturization efficiency (i.e. reduction in size while retaining the desired characteristics) obtained from experiments turned out to be worse than that obtained from numerical simulation. This is due to both manufacturing tolerances and possible differences in the substrate parameters.

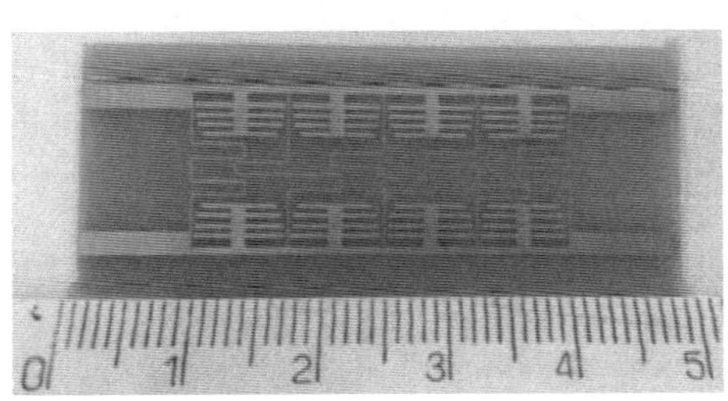

Figure 9. Manufactured prototype of the compact 3-dB coupler.

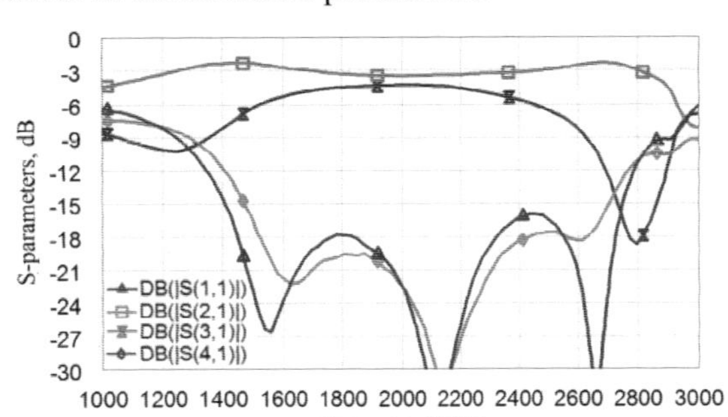

Figure 10. S-parameters of the manufactured compact 3-dB coupler.

IOP Conf. Series: Materials Science and Engineering **524** (2019) 012007 doi:10.1088/1757-899X/524/1/012007

Table 1. Designs comparison.

Parameters	*Standard*	*Compact (model)*	*Compact (layout)*
Bandwidth, MHz	1758	1375	1235
Relative bandwidth, %	87.9	68.75	61.75
Area, mm^2	2547	429.5	429.5
Relative area, %	100	16.85	16.85
Phase difference,°	90	89.3	88.5
Harmonic suppression	NO	YES	YES

3. Conclusion

In the present work we developed a miniaturized microstrip five-stub coupler tuned to the central operating frequency of 2000 MHz. The dimensions were reduced with the help of artificial transmission lines, whose lengths are smaller than the lengths of conventional quarter-wave segments. The placement of stubs in the free space inside the conventional layout made it possible to significantly reduce the area occupied by the coupler by 83.15%. At the same time, the operating frequency band was reduced by only 26.15%. Furthermore, a slight shift in the central operating frequency is observed with the designed coupler. This is probably due to the manufacturing tolerances or differences between the parameters of the substrates used for the simulation and the fabrication of the device. Further downsizing is possible by means of a more dense arrangement of the elements. However, this process leads to a further degradation in the frequency characteristics of the device.

Acknowledgement

The author is grateful to National Instruments for providing the simulation software NI-AWR Design Environment. The research was executed by the grant of the Ministry of Education and Science of the Russian Federation (project № 8.2538.2017/4.6). The research was performed using the equipment of the Urals Federal University common use center.

References

[1] Ye, X. F., Zheng, S. Y., & Pan, Y. M. (2016). A Compact Millimeter-Wave Patch Quadrature Coupler With a Wide Range of Coupling Coefficients. IEEE Microwave and Wireless Components Letters, 26(3), 165–167. doi:10.1109/lmwc.2016.2524581

[2] Phani Kumar, K. V., Barik, R. K., & Karthikeyan, S. S. (2016). A novel two section branch line coupler employing different transmission line techniques. AEU - International Journal of Electronics and Communications, 70(5), 738–742. doi:10.1016/j.aeue.2016.02.011.

[3] Lin, T.-W., Wu, J.-Y., & Kuo, J.-T. (2016). Compact filtering branch-line coupler with source-load coupling. 2016 IEEE International Workshop on Electromagnetics: Applications and Student Innovation Competition (iWEM). doi:10.1109/iwem.2016.7504934.

[4] Wang, Y., Ma, K., & Mou, S. (2016). A Compact Branch-Line Coupler Using Substrate Integrated Suspended Line Technology. IEEE Microwave and Wireless Components Letters, 26(2), 95–97. doi:10.1109/lmwc.2016.2517158

[5] Velidi, V. K., Subramanyam, A. V. G., Senthil Kumar, V., Mehta, Y., & Sanyal, S. (2016). Compact harmonic suppression branch-line coupler using signal-interference technique. 2016 Asia-Pacific Microwave Conference (APMC). doi:10.1109/apmc.2016.7931397

[6] Letavin Denis A., "Five-Stub Directional Device with Substantially Reduced Overall Dimensions," 2018 IEEE Radio and Antenna Days of the Indian Ocean (RADIO), pp. 1-2, Oct. 2018. DOI: 10.23919/RADIO.2018.8572386

RF Single-Pole Double-Throw Switch Based on Two-Port Memistor

J Vavra[1], J Bajer[2] and D Biolek[1,3]

[1] Department of Electrical Engineering, University of Defence, Kounicova 65, Brno, Czech Republic
[2] Department of Aviation Technology, University of Defence, Kounicova 65, Brno, Czech Republic
[3] Department of Microelectronics, Brno University of Technology, Technicka 10, Brno, Czech Republic

jiri.vavra@unob.cz

Abstract. The paper introduces a novel memristive two-port element for an efficient construction of Radio Frequency (RF) single-pole double-throw (SPDT) switches, the so-called Two-Port Memistor (TPM). In comparison with the classical Widrow memistor whose resistance depends on the charge passing through the auxiliary terminal, the resistance of the TPM is governed by the controlling signal applied to the programming gate. It is demonstrated that the SPDT switch can be constructed via two TPMs such that the RF chokes providing DC references and capacitors for AC couplings are no longer necessary. Since the TPM-based switches provide elegant and much simpler circuit solutions than the currently developed memristive switches, this paper could stimulate the research towards manufacturing the TPM as a solid-state device.

1. Introduction

The architecture of modern RF and microwave systems is rapidly changing. In addition to the conventional blocks such as the amplifiers, filters, mixers, impedance matching circuitry, limiters, phase shifters and others, the RF switches and other devices supporting flexible re-configuration of the signal processing chain play an important role. Typical examples are systems utilizing multiple antennas, Multiple-Input Multiple-Output (MIMO) systems, systems with diversity reception, etc. These devices need to be equipped with internal components that provide their mutual interconnection. Such components are typically based on solid-state RF switches. Another application where RF switches have become particularly important are T/R (Transmit/Receive) modules of radars and other devices utilizing AESA (Active Electronically Scanned Array) antennas, where the RF switch serves to select between the transmission and the reception signal path. These switches are usually based on Si PIN diodes, GaAs PIN diodes, GaAs MESFETs and HEMPTs, Si MOSFETs, SiGe HBTs, or GaN HEMPTs [1].

Recently, the RF switches employing the memristive devices arrived on the scene [2]-[5]. Memristors [6] or more general memristive systems [7] are two-terminal elements whose memristance depends on the internal state, which reflects the history of terminal voltage and current. The RF switch utilizing the memristive technology, called RFMS (RF Memristor Switch), profits from the memristor

nonvolatility and nano-scale. The low insertion loss, high isolation, and high cut-off frequency make the RFMS a perspective element for controlling future-generation RF applications.
A typical topology of the single-pole double-throw RFMS is shown in Fig. 1. The inductors L_{DC} provide DC grounding of the corresponding memristors M_1 and M_2, and the L_B arranges the DC path for driving both memristors via the controlling voltage V_{ctrl} which programs the M_1 and M_2 memristances to R_{on} and R_{off} for $V_{ctrl} > 0$ and to R_{off} and R_{on} for $V_{ctrl} < 0$. The capacitors provide the AC coupling.

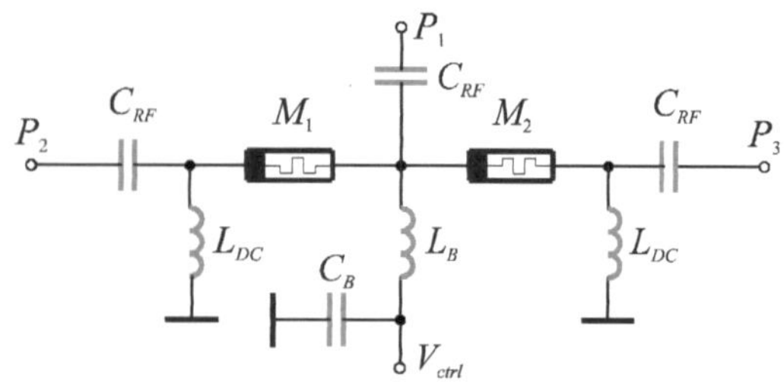

Figure 1. Conventional SPDT RF switch with two memristors [5].

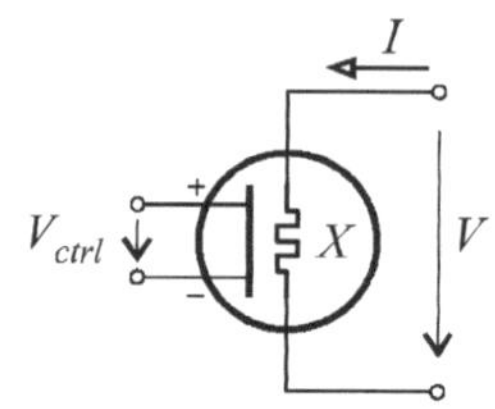

Figure 2. Schematic symbol of two-port memistor.

The goal is to replace the memristors in Fig. 1 with another type of device that will also provide the nonvolatile switching while concurrently enabling the removal of complicated DC/AC circuitries. The two-port memistors can be promising candidates for such nonvolatile switches.
This paper is an extended version of already published conference paper [8].

2. Two-Port Memistor

In accordance with the original definition of the memistor [9], [10], it behaves as a linear resistor whose resistance is controlled by the charge flowing to the auxiliary third high-impedance terminal. Since the ideal RF switch should also behave as a linear resistor programmed to R_{on} and R_{off} states, the memistor seems to be a better candidate for this role than the memristor whose resistance is modulated by the over-threshold signal. However, the current flowing to the controlling gate of the memistor requires a closed DC loop similar to classical memristor switching. The newly suggested two-port memistor with its schematic symbol as in Fig. 2 overcomes this problem by introducing one additional programming gate whose driving signal does not penetrate to the signal path. The voltage-current and state equations are

$$v = R(x)i\,,\ F(x, \dot{x}, v_{ctrl}) = 0 \tag{1}$$

where R is the memistor resistance, x is the internal state of the memistor, and F is a nonlinear function describing the dynamics of the process of programming the resistance.

It is important for the state variable governing the resistance to be programmed via a signal that is isolated from the resistive port. For the RF switch described in [2], the switch-on resistance is given by a conducting filament formed within the 35-nm air gap between two electrodes after applying a voltage ramp (0 to 3 V) across these electrodes. The state variable x in (1) reflects the geometry of this filament. Now it is necessary to apply a different mechanism for initiating the filament via another pair of electrodes. In any case, the corresponding model (1) of the related processes should be accordingly accommodated. For example, one may expect that the memistor state will be also modulated by the voltage or current of the resistive gate and that the resistance will also depend on these quantities. The latter means that the on-state switch would behave as an extended memristor. This observation is in accordance with the conclusions in [2].
In order to reflect the tunneling nature of the conductivity of MIM structures, the VTEAM model is used in [5] as a useful approximation of the more complex Pickett model [11]. For a more accurate

modeling of the on-state memistor, which can subsequently be used for small-signal transient and AC analyses, we used an approximation of the current-voltage Simmons equations from [12]:

$$i = \mathrm{sgn}(v)k_1 k_2^x \left[\sinh((k_3 + k_4 x)|v|) + k_5 (e^{k_6|v|} - 1) \right] \quad (2)$$

where k_1 to k_6 are the fitting parameters and x is the width of the tunnel barrier, the latter playing a similar role as the width of the conducting filament from [2].

3. Memistor-based SPDT switch

Two TPMs can be interconnected as shown in Fig. 3 such that the resulting SPDT switch does not require any additional AC/DC circuitries. The programming ports are in anti-parallel configuration, but they can be easily reconfigured according to the prospective demands for an independent programming.

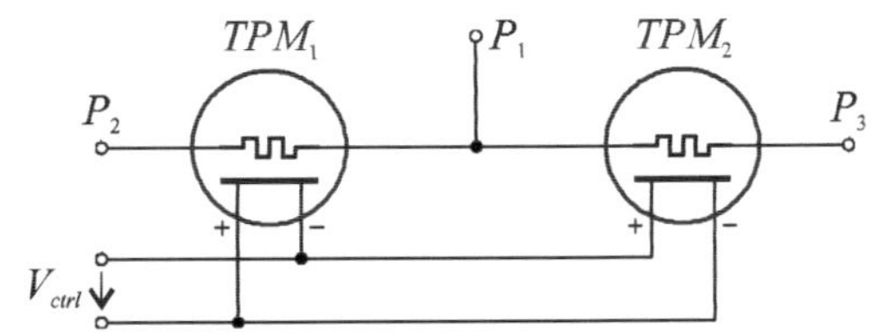

Figure 3. Circuit diagram of proposed memistor-based SPDT switch.

The SPICE simulation of the SPDT switch from Fig. 3 was performed with the TPM model derived from the RF lumped model of the nanoscale memristive switch published in [5]. Note that this model describes the small-signal behavior of the switch specimen from [2]. The corresponding equivalent circuit in Fig. 4 models only the resistive, not the programming port of the two-port memistor.

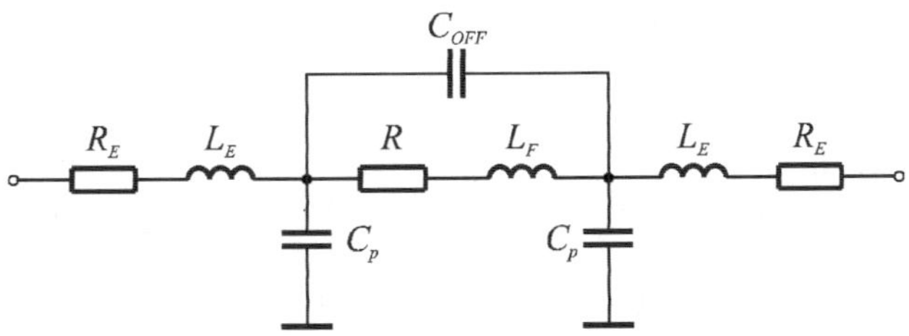

Figure 4. Equivalent model of the resistive part of two-port memistor [5].

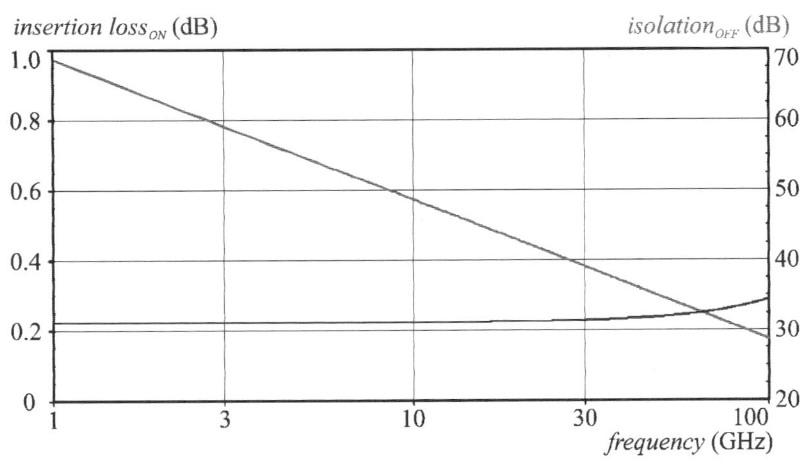

Figure 5. Simulated insertion loss and isolation of the TPM-based SPDT switch.

For the simulation, all the parameters in Fig. 4 except for the R (memistor resistance) were obtained from [5]. The resistive part of the memistor was modeled via the nonlinear equation (2), whose parameters fitted the dynamics of the TiO_2 memristor [12], with the R_{on} and R_{off} values specified in [2]. Fig. 5 shows the frequency characteristics of insertion loss and isolation of the switch which are in accordance with results in [5]. The blue line represents the insertion loss induced by an ON-state TPM_1 connected between the P_1 and P_2 terminals. The red one gives the isolation between the P_1 and P_3 terminals, when the OFF-state TPM_2 is connected to P_3. The insert loss and isolation at 30 GHz are approx. 0.25 dB and 40 dB, respectively.

4. Conclusion

A new circuit idea of nonvolatile RF switching is proposed. The two-terminal memistor is used for constructing the single-pole double-throw RF switch, which combines the advantages of the recently published memristor-based and the traditional FET RF switches. The main benefits of the proposed solution are simple circuitry, reduced power consumption due to the non-volatility of memristive devices, and simple switch programming via auxiliary ports located beyond the RF signal path.

To model the above memistor, the approximation was used of the current-voltage Simmons equations, which reflect the tunneling nature of the conductivity of MIM structures. The SPICE simulations are in accordance with the hitherto published results.

5. References

[1] Bahl I J 2014 *Control Components Using Si, GaAs, and GaN Technologies* (Boston: Artech House) p 310

[2] Pi S, Ghadiri-Sadrabadi M, Bardin J C and Xia Q 2015 Nanoscale Memristive Radiofrequency Switches *Nature Communications* **6** pp 1–9

[3] Nessel J A, Lee R Q, Mueller C H, Kozicki M N, Ren M and Morse J 2008 A Novel Nanoionics-based Switch for Microwave Applications *Proc. Int. Microwave Symp. Dig. (Atlanta)* pp 1050–54

[4] Vena A, Perret E, Tedjini S, Vallée C, Gonon P and Mannequin C 2012 A Fully Passive RF Switch based on Nanometric Conductive Bridge *Proc. Int. Microwave Symp. Dig. (Montrelal)* pp 1–3

[5] Wainstein N and Kvatinsky S 2018 A Lumped RF Model for Nanoscale Memristive Devices and Nonvolatile Single-Pole Double-Throw Switches *IEEE Trans. on Nanotechnology* **17** pp 873–83

[6] Chua L O 1971 Memristor–The Missing Circuit Element *IEEE Trans. Circuit Theory* **CT-18** pp 507–19

[7] Chua L O and Sung Mo Kang 1976 Memristive devices and systems *Proceedings of the IEEE* **64** pp 209–23

[8] Vavra J, Bajer J and Biolek D 2018 RF Single-Pole Double-Throw Switch Based on Memistor *2018 IEEE Radio and Antenna Days of the Indian Ocean (RADIO)* (Grand Port) pp 1-2

[9] Widrow B 1960 *An adaptive "ADALINE" neuron using chemical memistors* Stanford Electronics Laboratories Technical Report No. 1553-2 p 23

[10] Adhikari S P and Kim H 2012 Why are memristor and memistor different devices? *IEEE Trans. Circ. Syst.* **59** pp 2611–8

[11] Pickett M D, Strukov D B, Borghetti J L, Yang J J, Snider G S, Stewart D R and Williams R S 2009 Switching dynamics in titanium dioxide memristive devices *Journal of Applied Physics* **106** p 074508

[12] Biolek D, Kolka Z, Biolkova V, Biolek Z, Potrebic M and Tosic D 2018 Modeling and simulation of large memristive networks *Int. J. Circ. Theor. Appl.* **46** pp 50-65

Acknowledgments

This work was supported by the Project of Specific Research at K217, Research of sensor and control systems to achieve battlefield information superiority, UD Brno, and also by the Czech Science Foundation under grant no. 18-21608S.

IOP Conf. Series: Materials Science and Engineering **524** (2019) 012009 doi:10.1088/1757-899X/524/1/012009

Functional MRI and the Lagrangian description of flow

S L Vesely[1], S R Dolci[2], C A Dolci[3]

[1] ITB-CNR, Milan, IT
[2] Via Bergognone 31, Milan, IT
[3] Politecnico, Milan, IT

E-mail: `sara.vesely@cnr.it`

Abstract. To investigate local neuronal activity changes and functional connectivity during cognitive and sensory stimuli, functional magnetic resonance imaging (fMRI) relies mainly on the haemodynamic response to brain activation and on measuring small fluctuations in the blood oxygenation levels. Since the basal amount of glucose and oxygen consumptions of the brain are much higher than the activity evoked by cognitive stimuli, the effect of spontaneous cerebral activity must be deduced to evaluate the fMRI responses to given cognitive tasks. We report that in our experiments with water flowing in pipes at an oblique angle with respect to the scan slice a diffuse response can be observed besides other flow artefacts, which can undermine fMRI interpretations. We point out that those artefacts are considered as such because hydrodynamic models disregard the appearance of flow in the images, and more generally the differences between mechanical representations and images.

1. Introduction

Magnetic resonance imaging (MRI) scanners obtain pictures across living organisms that look much like anatomical sections without any need to resort to invasive techniques and exogenous contrast agents. The images are elicited mainly from tissue water, and are formed by sampling spatially encoded free induction decay (FID) responses to radio-frequency (RF) pulses. The contrast originating from the interconnected tissues with distinctive features is epitomized in a few macroscopic parameters. Among them the best known are the relaxation time of the magnetization *T1*, its dephasing time *T2*, and the water proton density ρ. Moreover, for clinical imaging purposes pulse sequences were devised, that are most sensitive to a single parameter, and accordingly the images obtained were called *T1*-, *T2*- and ρ -weighted.

Alongside with use in clinical diagnostics, MRI is applied to probe the organization of cerebral cortical areas since 1992 with the aim of linking together functional specializations and anatomical locations. Since the neuron action potentials known to occur from electrophysiology studies cannot be directly monitored with the MRI technique, the direct study of functional activation at neuron scale by functional MRI (fMRI) is precluded. The direct study of neuronal activity is ruled out for yet another reason. Although highly contrasted MR images across the human brain can be obtained, the achieved image resolution of about $0.2mm^3$ isn't enough to address the morphology of the nervous tissue – neurons and glia – at the cellular level. In fact, while under the light microscope stained anatomical preparations revealed the existence of a complex neuronal cytoarchitecture, and accordingly immunohistochemical and autoradiographic markers of cell lineages started to be combined with stereological methods in animal studies, there is yet scant evidence of how to relate MRI images of living human brains with the segregation of the physiological functions. This gap should be covered by MRI registrations of functional data. The current aim of fMRI is to uncover neural correlates of cognitive processes on a local basis, and explain them with the help of cytoarchitectonic maps. Our purpose is to draw attention via a simple flow phantom to the fact that a RF image across a living being doesn't coincide with the cut of its cellular structure model derived from an anatomy atlas.

2. MRI of flow

In 1950 Hahn showed that the *T2* relaxation parameter also depends on diffusion [1]. In 1951 Suryan confirmed that the signal intensity in nuclear magnetic resonance (NMR) is affected by flow conditions, and decreases with increased flow velocity along the main field under steady conditions, once signal saturation is avoided. In 1959 Singer applied the NMR technique to living beings, and later he showed that motion of water protons in a homogeneous magnetic field $\boldsymbol{B}_0$ can be accurately measured by time-of-flight methods and by phase encoding. On the threshold of the amazing MRI developments the design of pulse sequences apt to quantitatively evaluate blood flow in major blood vessels were taken up again [2].

In brain physiology liquid motion enters in two ways: first the diffusion of liquids between brain regions is an aspect of functional specialization, it being a clue for interconnectivity, and second hemodynamics supplies oxygen and glucose to an organ lacking any substantial capacity of energy storage. In fMRI, contrast is deemed to rely mainly on changes in the regional blood flow and blood oxygenation levels. As far back as 1881 an Italian physiologist hypothesized that the neural activity is tightly intertwined with precisely controlled changes in blood flow [3]. This insight is interpreted to mean that neural activation is coupled to the microcirculation that provides tissue perfusion.

In MRI, flow is a source of image contrast that allows extracting information on blood already from conventional spin echo (SE) and from gradient echo (GRE) sequences. Both sequences have drawbacks. Thus the former, SE, allows measurements in a restricted velocity range only, while the latter, GRE, is too sensible to field inhomogeneities $\boldsymbol{B}_1$ due to different tissue susceptibilities. The relationship between the two time constants is $\frac{1}{T2^*} = \frac{1}{T2} + \gamma \boldsymbol{B}_1$, with $\frac{\gamma}{2\pi} = 42.58MHz/T$. When deploying GRE pulse sequences to obtain $T2^*$ -weighted images, contrast regions appear around veins and in regions densely packed with venous capillaries that cannot be evidenced just as well while using a spin echo (SE) sequence. The supplemental contrast was ascribed to local susceptibility changes due to high deoxy-hemoglobin levels in blood. During oxygen uptake by red cells, hemoglobin switches from the paramagnetic to the diamagnetic state, and contrast disappears. The Lagrangian description of flow is assumed to be valid in numerical simulations, because the labeled water protons are precisely the particles displaced by motion [4, 5]. The link between local blood contrast and brain activation also depends on whether the course of the veins and capillaries correspond in the images and anatomically.

3. Materials and methods

Experiments were performed on a Tomikon MRI system using a $B_0 = 4.7T$ superconducting magnet with an inner bore diameter of $40cm$. Since $\omega_0 = \frac{\gamma}{2\pi} \times B_0$, the operating frequency was $\approx 200MHz$. The bore diameter of the inner probe head was $25cm$. The echo time $TE = 34.06ms$, by definition the time between the 90° pulse and the maximum of the echo, was factory set, while we put the pulse sequence repetition time $TR = 2s$. Sixteen sequential copies of a single slice were acquired by applying a 90° sinc shaped pulse envelope followed by sixteen 180° rephasing pulses (Carr-Purcel-Meiboom-Gill sequence – CPMG).

With reference to figure 1, a slice in a plane perpendicular to the magnet axis (z-axis) is frequency encoded by applying a slice-gradient G_s during the 90°-excitation. That FID is refocused sixteen times and its echo is acquired with read-gradient G_r on. Application of the latter gradient encodes a phase $\varphi \propto \gamma \int G_r(t)dt$ on the frequency response projected onto the x

IOP Conf. Series: Materials Science and Engineering **524** (2019) 012009 doi:10.1088/1757-899X/524/1/012009

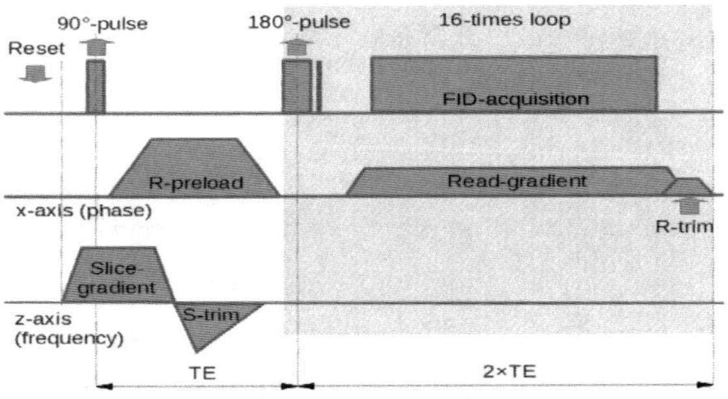

Figure 1: Spin-echo pulse sequence and encoding gradients. One 90° exciting pulse with z-slice gradient on is followed by 16 180° refocusing pulses. The echoed signals are sequentially projected on the x-axis for read out. Their Fourier (inverse) transformed FIDs show 16 times the profile of the image along the x-axis.

-axis, that is responsible for the shape of the profile in the transformed signal. The R-preload gradient compensates in advance for the FID dephasing introduced by the read-gradient. The R-trim aims at rephasing the FID within the loop by trial and error. In the method of image reconstruction from projections used here the section is reconstructed from data obtained by increasing the angle that the read-gradient axis forms in the xy plane before each new acquisition. As an alternative, the phase encoding method implemented as 2D-Fourier transform technique would change the dephasing of the preload gradient in the y-direction. Our acquisition matrices were 256×256, and the field of view (FOV) was $12cm$. Constant, laminar flow rates through a $3.5m$ long silicon tube of $0.9cm$ gauge passing four times through a rigid $9 \times 9 \times 5cm^3$ stand were obtained exploiting a Mariotte bottle as a tank. Nominal slice thicknesses were 4, 8 and $12mm$, and the measured flow rates Q were in the range $0.36 - 5.7cm^3/s$. Consequently, mean flow velocities Q/F ($F = \pi 0.45^2$) ranged from $0.57cm/s$ to $9.7cm/s$. Data for flow perpendicular to the slice, and inclined rearwards by 45° from the vertical (y-direction) were collected at the Fraunhofer Institut directed by prof. K. Gersonde. We refer to [6, 7] for further experimental details. For comparison, the lumen of cerebral vessels is seldom larger than a few millimeters, and the flow is always pulsed. Mean flow in large vessels is reported to be about a dozen centimeters per second.

4. Results and discussion

Since the 45° tilt has no component along the x-direction, flow doesn't affect the phase-encoding along it, and the imaged x-position of the tubing doesn't suffer any speed dependent overall displacements even without flow-compensating gradients [8]. As the velocity increases, though, the boundary layer of relatively stagnant flow near the wall is distorted along the y-direction, and so is the parabolic Poiseuille profile. The flow lines are cut, which is why the velocity component perpendicular to the slice is estimated with a higher error, and also depends on the slice thickness.

The $12mm$ slice is too large for our $9mm$ inner gauge tubing. In passing, with increasing slice size flow-induced field inhomogeneities contribute additional distortions to the image. The intensity of the odd echoes for the remaining slice thicknesses roughly satisfies an exponential curve $\boldsymbol{M}(t) = \boldsymbol{M}(0)\exp(-t/T2_{eff})$ with an effective time constant $\frac{1}{T2_{eff}} = \frac{1}{T2} + \frac{v}{d}$, where $T2$ refers to immovable water, and $\boldsymbol{v}/d$ accounts for the component of velocity perpendicular to the slice d of flowing water. The factory set TE value is too high, limiting the range of measurable velocities. Above $4cm/s$ the signal intensity for the $4mm$ slice at once reduces to $1/e \approx 0.368$, and prevents from estimating the value of $\boldsymbol{v}$. Again, our repetition time $TR = 2s$ is too short for the magnetization of water protons to return to the unexcited condition, because water has a $T1$ relaxation time in excess of $4s$. For this reason the SE pulse sequence keeps a steady state magnetization different from the fully relaxed one, and acquisitions occur under dynamic saturation conditions.

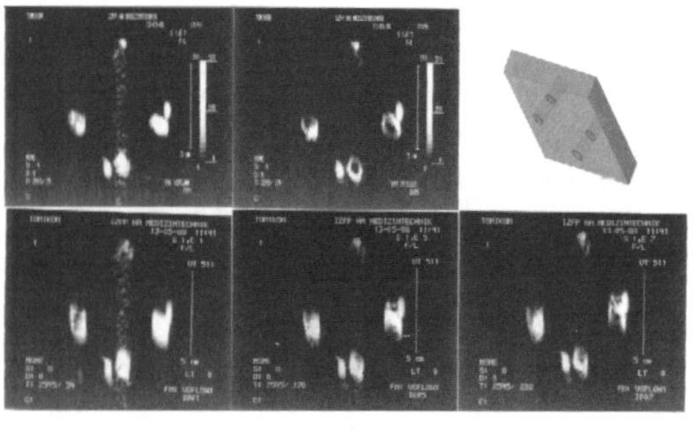

Figure 2: Images of water flowing at $4cm^3/s$ in a tubing oriented at 45° wrt. the slice. Top: 4mm slice, echoes 1 and 5. Bottom: 8mm slice, echoes 1, 5, and 7. Insert: the stand.

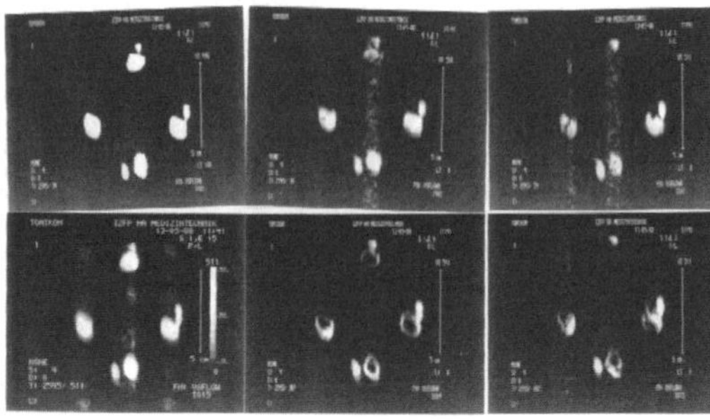

Figure 3: Water flowing in the 4mm slice at 45° wrt. it. Rising flow rates, 0.36, 2.6, and $5.7cm^3/s$. Top: 1st echoes. Bottom: echoes 15, 9, and 3.

In figure 2 the $4mm$ and $8mm$ slices are shown for $\boldsymbol{v} \approx 6.4cm/s$. The small bright spots appearing nearby are water filled reference vials parallel to the stand. With the stand at 45° the $8mm$ slice isn't entirely contained within it, but in no case the course of the real tubing just outside it can run parallelly to the slice. Yet, the first echoes show partially excited liquid flowing within the tubing, which is washed out in subsequent echoes. The residual excitation belongs to the response to the 90° sinc pulses other than the actual one. It hasn't yet decayed after TR, and is correctly refocused. The effect is further highlighted in figure 3. At low velocity stagnancy supports formation of refocused spots along the tubing, and contributes to the image at late times. On the opposite, rapid flow smears the course of the tubing already in the first image, affecting a wider range as the velocity rises.

5. Conclusions

Inflow and outflow from an imaging section is known to affect the intensity of a blood vessels' lumen. It is also known that flow oblique to the read-out direction causes a speed dependent displacement of the lumen from the vessel wall. Since our flow phantom with adjustable constant flow is simple, and we use a standard spin-echo sequence, we can show how the signal echoed from water flowing into sloped pipes, whose course lies manifestly outside of the imaged slice, is projected onto it. Velocity dependent smeared patterns appear on the image. When using MRI for demanding applications, like image-guided surgery, geometric and flow artefacts, that also depend on the chosen acquisition sequence, ought to be noticed and corrected as far as possible on each patient. Perhaps such smeared patterns also matter in connection with diffusion and capillary flow in functional MRI (fMRI). Anyway, MRI images look more like an electromagnetic field than an idealized mechanical cut, and cannot be forced to correspond point by point with anatomical sections. In particular, due to the above flow and diffusion *effects* on the images, the Lagrangian specification of spin flow may be apt to lend to misinterpretation if one is looking after small signal differences.

6. References

[1] Hahn E L 1950 *Phys. Rev.* **80** 580–94
[2] Axel L 1984 *Am. J. Roentgenol.* **143** 1157–66
[3] Sandrone S *et al.* 2014 *Brain* **137** 621–33
[4] Shadden S C and Arzani A 2015 *Ann. Biomed. Eng.* **43** 41–58
[5] Fortin A, Salmon S, Baruthio J, Delbany M and Durand E 2018 *Magn. Reson. Med.* **80** 1655–65
[6] Vesely S L, Dolci S R and Dolci C A 2018 *IEEE RADIO 2018, 15th–18th October* (Wolmar, Mauritius)
[7] Vesely S and Staemmler M 1989 *Rendiconti e Atti dell'Academia di Scienze Mediche e Chirurgiche* **143** 159–76
[8] Larson T C, Kelly W M, Ehman R L and Wehrli F W 1990 *Am. J. Neuroradiol.* **11** 1041–8